# The Unexpected Genius of Pigs

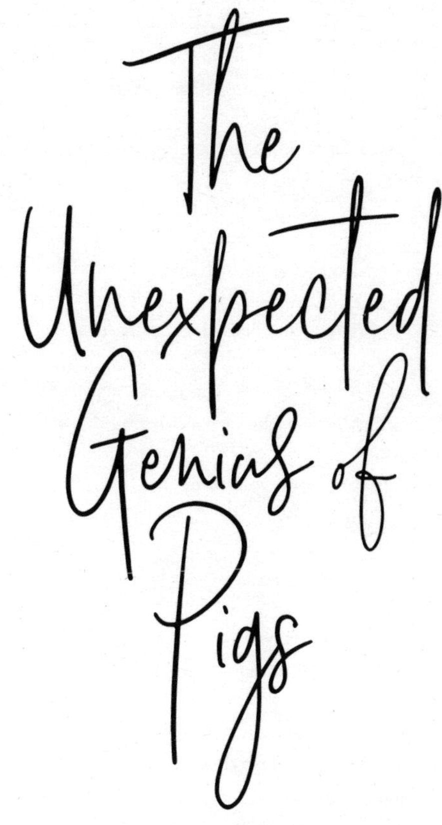


Matt Whyman

HarperCollins*Publishers*

HarperCollins*Publishers*
1 London Bridge Street
London SE1 9GF

www.harpercollins.co.uk

First published by HarperCollins*Publishers* 2018

1 3 5 7 9 10 8 6 4 2



Illustrations by Micaela Alcaino



A catalogue record of this book is
available from the British Library

ISBN 978-0-00-830122-4

Printed and bound in Great Britain by
CPI Group (UK) Ltd, Croydon

This book is dedicated to my dad

# Contents


| | | |
|---|---|---|
| 1. | **The Reluctant Pig-Keeper** | 1 |
| | A simple lesson | 1 |
| | Life before pigs | 2 |
| | Butch and Roxi | 3 |
| | An animal of mass distraction | 5 |
| | The growing presence of pigs | 6 |
| | Size and spirit | 8 |
| | Where there's muck … | 10 |
| | The unexpected genius of pigs | 12 |
| 2. | **The Ancestral Pig** | 15 |
| | Windows to the soul | 15 |
| | A pig in time | 15 |
| | The origin of the species | 17 |
| | The crossing | 20 |
| | At one with the pig | 21 |
| 3. | **The Mind of a Pig** | 24 |
| | Strands | 24 |
| | Survival of the smartest | 25 |
| | The pig in the labyrinth | 28 |
| | Wendy's world | 31 |

After Bertie 33
Egg heads 34
The journey and the destination 36
Immovable objects 39
Be more pig 41

**4. The Heart and Soul of a Pig** 43
Holly and Poddy 43
The matrilineal group 46
In the wild and on the farm 48
Mix, match and move on 49
Protection racket 50
The reporter in the pigpen 51
Social security 54
The complete pig 55
Meet the Swedes 57

**5. The Language of Pigs** 62
Rocky 62
Conversations with a pig 63
'I am here' 63
Sound and fury 65
Where there is hope … 66
The silence of the pigs 67
In a world of their own 69
Talking to the animals 74
Cilla 75

6. **The Pig's Snout** 78
Lost treasure 78
The tool of the trade 81
Beyond the boundary 85
Before the drop 86
After the rain 88
No secrets from a pig 89
The orchard next door 90

7. **The Realm of the Pig** 94
Come home 94
Home is where the food is 95
The trustworthy pig 96
Helga 98
The pastoral pig 99
Sleeping under the stars 100
Creature comforts 101
The clean pig 103
The mud myth 104
Bath time 107

8. **The Sow and the Boar** 110
The heat is on 110
Let's hear it for the boars 113
Herbie 116
The lesser boar 117
Love is in the air 120

**9. What Pigs Can Teach Us About Parenting** 123
In pig 123
The nesting instinct 123
A magpie in the house 127
The mothering instinct 129
Smart suckling 131
The other mothers 133
The spare-part parent 135
Sybil 136
Striking out 137

**10. The Companion Pig** 139
Not just for Christmas 139
Man's beast friend 142
Pastures new 145

Acknowledgements 149

# 1

# The Reluctant Pig-Keeper

## A simple lesson

Keeping pigs taught me a great deal about myself, and very little about the animals in my care. In the years that Butch and Roxi were part of my family, I discovered that my patience could be stretched almost limitlessly. I also realised that things I had considered to be important didn't really matter, like flowerbeds and much of the fencing surrounding the garden. As a father of four young children, I was no stranger to hard work and responsibility. Even so, no amount of nappy changing could have prepared me for the muck I faced on a daily basis. The experience brought me closer to my wife, Emma, in the never-ending challenges presented by our porcine pair, but not once did we give up on them.

Above all, for all the trials, escape bids and destruction, I learned about love.

## Life before pigs

Looking back, I have only myself to blame. We live in the West Sussex countryside, in a brick and tile house on the edge of woods. There is a garden where our children used to like to play, and neighbours on each side. For some time, I'd kept chickens in an enclosure at the back. The area was defined by a picket fence that crossed behind a small apple tree and attached to the front corner of the shed. In effect, it was a paradise for poultry. My six-strong posse poked and scratched about in an abundance of space, and always sailed to the gate to greet me whenever I wandered down to see them.

When a fox attack put paid to all but one of my flock, it prompted me to ask what animal might deter a repeat visit. What I had in mind was something that would send out a clear signal, like a crocodile, a pool of piranhas or an angry bull. I wasn't being serious when I suggested a pig, though I'd heard they often spooked foxes. For Emma, it was reason enough to go online and do some research. When she found a type that could supposedly snuggle inside a handbag, it was a done deal.

'These aren't normal pigs,' she pitched to me. 'They're *mini*pigs.

To be fair to Emma, she had done her homework. It's just that at the time this amounted to trawling through a raft of irresistible pictures of impossibly small pigs in baby booties, and scant hard facts about what set them apart from your everyday swine. All she could do was take the word of the few breeders that she found who specialised in minipigs. According to them, pint-sized porkers grew just 12 inches high, which is roughly the same as a Terrier. They were smart, child-friendly, easily trained and happy to live under the same roof as us.

Emma did tell me a lot more about them, but I had stopped paying attention when she delivered the clincher by assuring me I'd barely notice them. By then, my family were totally sold. A run-of-the-mill piglet costs about £30. For an eight-week-old minipig, you're looking at anything between £500 and £1,000. Despite the hit, Emma believed it would be an investment. 'The children will remember this,' she said. Looking back, she wasn't wrong. It's just that I don't think the experience shaped their lives in the way she had hoped.

## Butch and Roxi

The new arrivals pitched up in a cat basket. In a bid to butter me up, perhaps, Emma nominated names I'd once proposed for two of our children only to have them dismissed out of hand. As per the breeder's sales pitch to Emma, they were no bigger than kittens. Perfectly pig-shaped and honking in a high pitch, my first thought was to check their bellies for battery compartments. They just seemed too good to be true. Over the course of their first weekend with us, the sibling pair were effectively magnets for the attention and affections of Emma and the children. As I worked from home, writing books in an office at the front of the house, I used the opportunity to slip away to the typeface.

Then Monday arrived. With the children at school, dropped there by my wife on her way to the office, the task of looking after Butch and Roxi fell to me.

From that moment on, the gap between pig-keeping fantasy and reality opened up like a chasm. From where I was sitting, in front of the computer trying to write for a living, they drove me to distraction. Thoughtfully, for the pigs at any rate, Emma had decided to locate their little ark in my office so that I could watch over them. In a sense, that's exactly what I did, spending more time peering over my shoulder at a string of interruptions than facing the screen.

Contrary to popular belief, pigs are hygienic creatures. They'll create a toilet as far from their sleeping quarters as they can. In our house, despite the litter tray Emma had installed in my office, that meant trotting into the front room and slipping behind the television in the corner. Noise-wise, they weren't too bad. In fact, the snuffling and grunting was really quite soothing as I worked. It was only when the phone rang that the atmosphere soured. It might have been something to do with the frequency of the ring-tone, or perhaps pigs just like a singsong. Whatever the case, Butch and Roxi would respond by squealing away. It's tough enough trying to come across as a professional in a home environment. Now, it sounded as though I was working out of a barnyard.

## An animal of mass distraction

Of course, everyone knows that taking on a young pet can be testing. Dogs need to learn you're the boss, while cats take a while to work out how to manipulate you to their advantage. Pigs are a lot like toddlers. They can be gentle and inquisitive souls and then break into a tantrum when things don't go their own way. Unlike little kids, as I found out, they don't grow out of this behaviour. Over time, it just becomes more forceful and out of place in a domestic environment.

What's more, there are strict rules and regulations to observe, as set out by the Department for the Environment, Food and Rural Affairs (DEFRA). In giving pigs any kind of food that's been in a kitchen, for example, I risked contravening various biosafety laws. It could earn a hefty fine, but our little livestock didn't know that. Nor did the youngest of my children as he toddled around with a biscuit in hand and two little pigs trailing after him like low-level jackals. Ultimately, you only have to witness a minipig having a meltdown because you won't share a sandwich to recognise that life might be easier for everyone if they moved outside.

Butch and Roxi lived inside with us for just a very short time. As the novelty wore off, it very quickly became clear to me that the house was no fit environment for a pig of any description. They're purpose-built to dig about in the soil, seeking out roots and buried treats, not jam their snouts into the wine rack or flop about in front of the TV waiting for the lottery results. Surprisingly, it didn't take much to convince Emma and the kids. While they had also come to recognise that this special breed of pig didn't require carpet under hoof and central heating, I think they also craved a little peace. To be sure they didn't change their minds, I adopted some ex-battery hens to befriend my sole surviving bird and then played the fox protection card.

And so it was, with a clutch of post-institutionalised chickens perched on the handle of the toolbox beside me, I converted the side of the shed into cosy sleeping quarters for Butch and Roxi. The fencing seemed sturdy enough, I decided, having given it a shake, and there was more than enough space for everyone to peacefully cohabitate.

## The growing presence of pigs

In the clear light of day, once the turf war between pig and poultry settled down, it became clear that Butch and Roxi were no longer quite so mini. Roxi developed the fastest. In fact, there was a period when she appeared to look bigger every single time I went out to serve up their beloved pig nuts for breakfast and supper. They also fattened themselves up somewhat by gorging on all the acorns that dropped from the oak tree, along with the leaves when they fell in the autumn.

While Roxi rivalled our late German Shepherd dog in size, Butch compensated by becoming stockier and transforming into a mighty excavator. Having taken over the chicken enclosure, the pair turned it into a cratered mess of mud. I felt so sorry for the birds that I would let them

out onto the lawn. Around that time, unwilling to miss out on the party, one of the pigs learned how to lift the latch on the gate. Lashing it shut kept them in check for a while. Butch and Roxi responded by growing big enough to prise away the picket fencing with their snouts.

Surveying the remains of the garden one day, as the pigs slept off their hard work inside the shed, I refused to be defeated. I set about strengthening the fencing – effectively an epic bodge job – and just assumed that our minipigs must have reached their full size. Which makes me laugh in retrospect.

## Size and spirit

As time passed, and friends or neighbours visited, they'd often catch their breath on seeing the honking great beasts amid the craters and spoil heaps that was our garden. Within a year, Roxi stood thigh-high to me and had developed a taste for house bricks. She kept rooting them up out of nowhere and then crunching them into powder. A pink pig with dark splodges, she had bat-like ears and a face that could best be described as 'shovel-like'. She was densely built as well; a solid mass of muscle, fat and obstinacy. Had we let her stay in the house as a piglet, we'd have needed a winch to get her out.

Butch wasn't quite so monstrously big. In the right light he could even have passed as cute. He was all black with an elongated belly and a soulful expression modelled on Yoda from *Star Wars*. Castrated at an early age, because frankly, the consequences of leaving him intact were unthinkable, our male minipig also reminded me of a henpecked husband around Roxi. She really did rule the roost, much to the displeasure of the chickens. Had she taken to crowing at sunrise, I don't think any of us would have been surprised.

Without a doubt, it was a struggle to serve the growing needs of our little livestock. The bolstered picket fencing felt like a dam containing rising waters, but it held all the same. I can't be as positive about the six-foot close-board garden fencing that formed the back of the enclosure. I panicked the first time I found a splintered, pig-shaped hole in it one morning, and spent the whole day tracking them down. The second and third time was equally troubling. When it happened again I began to wonder if they had been sent on purpose to test the boundaries of my patience.

Around this time, Emma took it upon herself to contact the breeder. Butch and Roxi didn't exactly match the pictures on the website of cute little creatures curled up in a shoe box, and so she reached out to address it with them like a consumer's champion crossed with an avenging angel. I have no doubt that my wife would have taken them to task in a reasonable manner, while leaving them in no doubt that passing off pigs in this way was something that had to stop unless they wanted a tall and angry blonde on their doorstep. As it turned out, I can only think that another disgruntled minipig owner had got in before her, because the breeder was no longer trading.

Even when Butch and Roxi behaved themselves, there was no ignoring their ever-increasing size. Despite the squealing, and the fact that our garden looked like a battleground, our neighbours were surprisingly understanding. I lost count of the number of times I had to pre-empt a noise complaint by popping round to apologise. I ended up giving away all the eggs produced by our chickens by way of compensation. In conversation about our plight, they seemed to recognise that we had no idea what we'd let ourselves in for. I dare say they quietly considered us to be foolhardy and impulsive in falling for the idea of keeping pigs as pets without due diligence, and they'd be right.

In some ways, however, we were lucky. Despite the sacrifices, we just about had the space to serve Butch and Roxi's welfare. Their upkeep dominated our lives. I even called a halt to my work as a novelist to write a cautionary tale about the experience in the form of a memoir. So, what happened here? Had we been conned?

While belatedly attending a pig-keeping course, a conversation with the wise old boy running it opened my eyes to the reality of our situation. He believed the long-standing interest in pigs, and the money that the idea of a miniaturised version could command, led some people in the business to cut corners. 'Minipigs aren't a recognised registered breed,' he told me. 'Anyone can mate two small-sized pigs, but there's no guarantee that the offspring will stay small. That would take generations of strictly controlled breeding. Maybe we'll see such a thing in thirty or forty years from now,' he added, though it offered little comfort. 'But what you have are two mixed-breed pigs.' As for their status as brother and sister, the man took one look at the photograph I showed him and chuckled to himself.

So the minipigs were a myth, it seemed. Piglets passed off for profit. A unicorn for our age, or perhaps just for people like us who were drawn to the idea of a pig as a pet. Yes, small breed pigs exist, like the Vietnamese pot belly and the kunekune, but much depends on your concept of size. The idea of any adult pig that could fit inside a handbag is nonsense. In fact, an adult pig could have that kind of thing for lunch if you left it lying around. The fact was Butch and Roxi were two very expensive bog-standard mongrels. But despite it all, we cared for them deeply. In a way their presence served to bring us closer together as a family under fire.

Now, we're not the kind of people who would ever give up on their pets. It was hard work, but Emma and I learned a great deal about responsible pig-keeping, and that has its own rewards. Winston Churchill once observed that in looking a pig in the eye you will find an equal peering back at you. I'm not so sure. Those times I levelled with Butch and Roxi, usually in pleading with them to just give me one day without grief, I found two grunting creatures meeting my gaze with more lust for life and sheer determination than I could ever muster. It was also a bonding experience. We were in this together, man and pig. Throughout, my wife and I always wanted to do the right thing for them. We had been ill-prepared and enchanted in equal measure. And yet no matter how much they tested us, Butch and Roxi's welfare was always our priority.

I can console myself a little by the fact that we weren't alone in falling for the minipig myth. Other households had invited them into their homes with the very best of intentions, only to find they'd outgrow their welcome. Across the country, animal sanctuaries began to accommodate pigs that were as large as they were lonely and sad, which was the last thing we wanted to do. Our pigs were a part of our lives, even if they had come to dominate every aspect of it. Emma and I agreed that Butch and Roxi had just as much right as us to a happy and fulfilling existence, and we pledged to do everything we could to furnish that for them.

## The unexpected genius of pigs

A long time has passed since I called myself a pig-keeper. The emotional scars have healed and the grass has come back with a vengeance, thanks to all that compost. I can look back on this episode in our lives with fond memories, and even smile to myself at some of the escapades that left me seething at the time. As well as encouraging us to give up eating meat, thanks to a heightened respect for animals and their welfare, it's left me with a fascination about what makes pigs tick. We invited a pair into our world and they trashed it, but what's it like in *their* world?

Now that I have pig-free time and space to think, I'm interested in finding out more about life through their eyes. I have no doubt, from my face off with a pair who had my measure from the start, that this is a species with hidden depths. I'm not suggesting a pig has a penchant for algebra, painting or poetry, but there's something extraordinary going on between those ears that I'm keen to explore. In some ways, I think, it's a perfect storm of instinct and intelligence that means when a pig puts its mind to something it always gets what it wants.

Pigs aren't just smart, they're also strikingly sociable. Butch and Roxi were inseparable and, though not siblings, were they companions by necessity or genuine soulmates? Roxi would regularly use her size and heft advantage to shove Butch from the feeding trough, while our boar was quicker on his trotters and could scurry away with an apple in his jaws before she could catch up. So, they jostled for food and yet served as comfort blankets for each other at night in a kind of intimate snout-to-backside arrangement.

As for escaping and running away, I could guarantee that no matter where Butch and Roxi ended up, I would always find them together. So do pigs form loyal friendships as we do, or undertake feuds with one another? And what's with the need for a partner in crime? Can they love and loathe, offer comfort, or share wisdom and advice? Are they playful or mischievous, truly lazy or actually greedy, as we often say when suggesting someone is behaving like a pig? Free from the technology that links us, how do they communicate and what do they say? And what is it that drives them

to dig from dawn to dusk in order to unearth a single acorn? It's all a mystery to me, but one I'd like to pick apart in a spirit of curiosity and enthusiasm in order to understand them better.

With help from people who have looked a pig in the eye far deeper than I ever managed, I intend to learn more than I did from all the mistakes I made as a reluctant keeper. Not just about pigs and their personalities – and we'll meet quite a few – but what it means to connect with these animals and to recognise that they lead lives that can be just as complex, challenging and rewarding as our own.

# 2
# The Ancestral Pig

## Windows to the soul

Looking a pig in the eye, as Churchill famously discovered, can be an unnerving experience. Levelling with a dog in the same way will instinctively tell you that you're in charge, while most cats simply turn away dismissively, but a pig prompts pause for thought. Nose to snout, gazing into those little orbs you'll find a depth of contemplation to match your own. A pig will blink just like you, batting eyelashes like the wings of a resting butterfly, and invite you to glimpse a spirit as shining and sophisticated as your own.

They'll also grunt, in a primal way, as if to remind you of their origins.

## A pig in time

The ancestral line from today's domestic pig dates back between nine and 13,000 years to the European Wild Boar. Still in existence today, these are powerful, bristle-backed beasts, long in the skull and often dark in colouring compared to their pink and hairless cousins. They stand on broad shoulders that taper towards their hind legs, like a large breed dog in a heavyweight division.

Also known as *Sus scrofa*, variations of the wild boar can now be found from Africa to Asia, the Far East to Australia, and in a variety of habitats, including forest, scrub and swampland.

They live in groups and move around depending on what resources are available to them. For a variety of reasons, wild boars are drawn to areas of dense vegetation. In short, their world revolves around three elements: food, water and protection. They can find this in undergrowth and beneath leaf canopies near rivers and streams, but it's also something humans can provide – which is where the connection between us was first forged.

In a bid to find out more about what drew the boar into our world, I visited Michael Mendl, Professor of Animal Behaviour and Welfare, a recognised expert in the cognition, emotion, individuality and social behaviour of domestic animals. He's also a man who is passionate about pigs. A warm, softly spoken and thoroughly engaging individual, Professor Mendl opens the door to his office at the Bristol Veterinary School wearing a T-shirt and jeans. This is my kind of academic. And when he gets on to our chosen subject, I am delighted to learn that his immense knowledge base also comes with recognition that we can never know precisely what makes another animal tick – and that within this mystery lies some magic.

'We can look at fossil records from human settlements dating back about 10,000 years ago for signs that wild boar were on their way towards domestication,' the Professor begins. 'It's likely that an association developed in terms of co-location, and that the boar ventured into the settlements because they were scavenging. As omnivores there would be things that they were interested in, like food that people had left. And I imagine those people looked at the boar and had ideas of their own,' he adds with a grin. 'There's plenty of information available to help us calculate when this happened, but the fact is we don't really know *how* the pig was first domesticated.' The Professor considers me through his glasses for a moment. 'You can always make up a story about it.'

## The origin of the species

By and large, a pig is a docile and benign beast. There are times when it's best to steer clear, as we'll discover, but with some understanding they're generally easy to read. Instinctively, I think, before reaching out to scratch a flank or rub behind an ear, we'll talk to them. There is something about pigs that always prompts us to do this. They respond to human voices, just as we respond to them. We speak very different languages, but the tone of both a grunt and our greeting seems to effortlessly cross the divide.

The wild boar, on the other hand, is a very different creature. Visiting the outskirts of Bucharest on a work trip recently, I decided to use some free time to go for a run. I have always been a runner (apart from the period when Butch and Roxi devoured my time). I find it helps to clear my head if I've been writing all day, and basically serves both my mental and physical health. That day, unfamiliar with the city, I planned a route on Google Maps. I had assumed I could plot a circuit that might take in a park with a lake and suchlike. I just hadn't realised that this quarter of the Romanian capital contained a vast swathe of dense forest. From above, this dark and ragged expanse looked completely out of keeping with the grid of avenues surrounding it, one of which I would have to follow to pick up the trail path. Aware that stray dogs roamed the streets, I asked at the hotel reception if it would be wise.

'No problem,' the receptionist told me. 'The dogs are harmless if you leave them alone, but in the forest you must watch out for the boar.' Her English, and my non-existent Romanian, didn't allow for finding out more. I just thanked her, smiled, and headed outside, where exhaust-stained snow had been shovelled to each side of the avenue.

Despite being dressed in technical shorts and a luminous Lycra top, with the receptionist's warning in mind I felt like an age-old character from *Grimms' Fairy Tales*. All the way to the forest, no more than a mile at most, I dwelled on what I might face. I passed lone street dogs that paid me no attention, and a Rottweiler behind a fence that chased alongside me for a while. It was noisy, but I wasn't alarmed. We see dogs of every temperament. They live among us, unlike the animal I had been warned about, and as I approached the trailhead I felt as if I was leaving my world and entering one that belonged to them.

I have never seen a wild boar for real. I know they're beginning to populate pockets of the UK once more, but I still think of them as a livestock version of the Loch Ness

Monster. Having become hunted into extinction in the seventeenth century, their quiet reappearance in forested regions from Scotland to the south coast of England is largely believed to have begun in the 1980s as a result of escapes from captivity. Today, it's estimated that 4,000 wild boar could be at large in the British countryside. It may not sound like many, but an adult male can weigh in at 150kg of muscle and tusk, and is unlikely to turn tail if startled, in the manner of a rabbit or deer. In rural parts of Europe, however, especially to the East, the wild boar is commonplace, and this was uppermost in my mind on leaving the avenue behind and heading deep into the trail.

With my running shoes crunching through the snow and the low sun hanging behind the trees, I found myself becoming all eyes and ears. The only thing I knew about wild boar was that as territorial creatures they could be aggressive when disturbed, and here I was breezing through their kingdom without a pass. I admit to feeling some apprehension, seeing movement in the thickets when there was none, and I picked up my pace along with my heart rate on registering the sound of something scramble away. When I heard a distant but guttural snort my nerve deserted me completely. In my mind, I faced imminent attack by a beast that suddenly embodied my greatest fears. As casually as I could, I turned and ran back the way I had come.

'You were lucky,' the receptionist told me when I reported the experience on my return. I am pretty sure she was simply telling me what I needed to hear. There was every chance that I had just been startled by my own shadow, but as a hotel guest I hadn't paid to be ridiculed. Nevertheless,

I returned to my room with a renewed sense that we are hardwired to be wary of wild boar. Like the bear, it's a creature that we consider to exist across a divide – one that represents danger, should we venture far from home.

## The crossing

With no nice hotels to hide out in, or room service to cater for their needs, our ancestors were right to be wary of the bear and the boar when they ventured into the woods and forests. After all, these creatures had a significant advantage in their domain: they would be aware of your presence before you saw them, which would be sure to unsettle anyone but the hunter. So they were best left alone.

And yet the wild boar viewed the world beyond their own through different eyes. Unlike the bear, the boar ventured out from their kingdom and into ours. By extension, they duly broke the spell between man and beast. I like to think they did so with some trepidation, crossing the line under the cover of night to claim the scraps that had been discarded on the outskirts. In a sense, they had found a way into human life that presented no threat. If anything, by clearing the ground of waste that would otherwise attract vermin, these pioneering forebears of the common pig offered something back, and thereby laid the blueprint for a relationship that would thrive.

'The boar really is quite a wild animal,' Professor Mendl points out when considering how we took things to another level on discovering a taste for the meat. 'Some would have been bolder than others, and willing to interact with people, and so the selection and breeding process would have been gradual.'

## At one with the pig

Studying fossils, it's possible to look back through time and see the pig evolve as humans moved from foraging to farming. For the swineherds through the ages, however, the emergence of the docile beast from its wild ancestor would have been imperceptible. Every generation continued the work of the last, slowly shaping form and nature through one century and on to the next. The tail coiled as the skull broadened and the nose flattened into a snout. The dark bristles softened and yielded to a pink and hairless skin, while the ferocity and fury that defined a wild boar under fire burned out to reveal the gentle soul we recognise today.

In many ways, the pig allowed itself to become domesticated in order to earn a place in our world. In changing itself for ever, and submitting to our needs, it brought us closer together.

Throughout the ages, our relationship has become ever more tightly intertwined. The pig assumes the final position in the Chinese zodiac, having shown up last when Jade the Emperor called a gathering of animals. In this story, the pig is celebrated for its honesty and determination, having admitted it fell asleep along the way, and yet it's believed this might also be where it picked up a reputation for being lazy.

Other areas of folklore see the pig ascribed with different qualities. In Ancient Egypt, the pig was associated with Set, god of storms and disorder, and by Native Americans as a herald of rain; while the Celts considered it to be an icon of fertility and abundance. Pigs have impacted on religion, most notably in being unfit for consumption under laws of both Islam and Judaism. Buddhism portrays a deity called Marici as a beautiful woman in the lotus position astride seven sows, and the New Testament tells the story of the exorcism of the Gerasene demoniac, in which Jesus cured a man possessed by casting the evil spirits into a herd of swine.

Across the world, from one culture to another, pigs have come to represent extremes of the human spirit, from sloth, gluttony and dirtiness to irrepressibility and sheer lust for life. There is no middle ground. Love or loathe the common pig, it has made its presence known.

Today, it is believed that the world pig population exceeds one billion. The vast majority are raised for meat on an industrial scale, with breeds such as the Large White and Land Race, Duroc and Piétrain optimised for fast growth and large litters. While pig farming became big business after the Second World War, rare breeds have seen a renaissance in recent decades. Whether driven by welfare issues or the texture and taste of the meat from pre-industrialised breeds, there's something reassuring about the sight of an old-time pig turning the soil under the sun.

Visit any smallholding or select farm and you'll find the Gloucester Old Spot, the Berkshire and the Tamworth, the Oxford Sandy and Black, and the Saddleback. While these breeds are physically distinct from one another, with some showing behavioural traits such as the Tamworth's remarkable escape skills, you'll discover that each pig also possesses a force of character and spirit that immediately defines one from another.

Approach a pen or a field of pigs and you can be guaranteed a greeting. Bold or shy, they'll always register your presence – especially if you bring something to eat – and never cease talking to you. From *The Three Little Pigs* to George Orwell's revolutionary swine, Winnie-the-Pooh's timid friend, Piglet, to Miss Piggy, Wilbur and Babe, we have anthropomorphised pigs in the stories we tell one another to better understand ourselves.

Pigs are far from human, however. With outsized ears and disc-shaped snouts there is something unearthly about them, and yet eye to eye, that connection with us is there. What goes through their minds is something we can only wonder at. Be it driven by emotion, instinct or a blend of both, the bond we share has strengthened over time and continues to grow. Just look at advances in modern medicine. Not only has the pig genome been sequenced, opening up their inner world, we have established that our anatomical and physiological make-up – including our cardiovascular systems – are remarkably similar. While we already call upon pig tissue in some life-saving surgical procedures, there will surely come a time in the near future when the pig becomes a viable donor for organ transplants.

In a sense, our hearts already beat as one.

# 3

# The Mind of a Pig

## Strands

The animal kingdom will always be a mystery to us. We can explore it in many illuminating ways, from a biological or behavioural perspective, but we will never share that singular strand that binds a species together. As humans, we understand each other in a way that the dog and the cow can only observe from their own world. In the same way, when we peer into the realm of the pig, we do so from a step away.

So, in asking ourselves what makes an animal like a pig tick, let's not be afraid to call upon our imaginations to bridge the gap. George Orwell did just that in *Animal Farm* by proposing that, 'being the cleverest', the organisation of the livestock into a force to be reckoned with should fall naturally to the swine. In the same way, we'll never truly know whether a pig feels love or grief, calculates, deliberates or daydreams, of course, but if we take a leap of faith and accept that it's as sentient a creature as we are then we have the vocabulary to explore that missing strand.

## Survival of the smartest

Butch, my not-so-minipig, lived in the shadow of his so-called sister. Growing up together, they had certainly bonded like siblings. They shared the same sleeping quarters, flopped side by side with their snouts poking out, rose at the same time as the cockerels and then picked and foraged their way through each day.

Size informs the porcine equivalent of a pecking order, and this dictated that Roxi was the dominant pig. At feeding time, she could shove Butch to one side with such force that it could knock him off his trotters. I found this alarming at first, and worried about a breakfast-related injury. I took to filling their big rubber feeding bowl and then trying to engineer things so that Butch could get there first. This involved standing between Roxi and the bowl. Then I found that Roxi would just try to barge *me* out of the way, so that strategy didn't last long.

Despite her insistence on breakfasting before Butch, she never finished the pig nuts I had measured into the bowl. This might have been down to the fact that Roxi couldn't manage double helpings, or perhaps she purposely left enough for him. Either way, Butch always got his breakfast. It's just he only ever did so on her terms. Until, that is, Butch began to use his brains.

To a certain extent, he was only following Roxi's example. She had figured out that by making a lot of noise from the moment she woke, I would come running like her personal servant. When I say noise, I mean a blood-curdling squeal that must come close to what would accompany the opening of the gates of hell. As she did so at the crack of dawn, it always forced me to career from the house in a half-tied dressing gown in a bid to shut her up before every resident along the lane turned against us.

Over time, it made me so anxious that I took to setting my alarm just ahead of her call to arms. That way, I would at least have time to slip on my wellington boots rather than bound there barefoot. And it worked, for a while. If I crept down to the pigpen, and lifted the latch without making a squeak, I could leave out breakfast and be back in bed before Roxi had a chance to rise and draw breath.

Several weeks into my new strategy, sleep deprived but with the peace of the neighbourhood intact, I found myself under observation as I quietly filled the bowl. I paused and glanced across to the pigs' sleeping quarters. In the breaking light, a pair of beady eyes peered back at me.

'Shhh,' I whispered at Butch, and finished the task at hand.

I retreated to the garden and closed the gate behind me. As I did so, the little black pig slipped out into the open so quietly that all I heard was the crackle of straw. With the sun just a promise behind the woods, he stretched and then crossed to the bowl. I fully expected Roxi to follow. Instead, as he began to pick and graze, she slumbered on with barely a twitch of her ears. It wasn't until I was back in bed, in fact, with my alarm reset for an extra half an hour, that I heard the familiar rumble and squeal. Only this time it stopped just as soon as it had started. In the silence that followed, curiosity got the better of me. I crossed to the window overlooking the garden, peeled back a curtain and peeped outside. There, in the first bars of sunshine, just as Roxi finished guzzling greedily on what looked like a fair share of pig nuts, I watched the cunning little boar make his way back to bed for a post-breakfast snooze.

As a one-off, I considered the moment worth sharing with my wife and kids. Over the course of the next few mornings, when I found Butch waiting for me beside his snoring partner, and then repeating the same trick, I marked him down as being as shrewd as he was small.

It took a while for Roxi to rumble him, and prime herself to wake up just as soon as Butch slipped from their bed. Naturally, she charged out and reclaimed her position as the

pig entitled to first pickings. Butch seemed resigned to the situation, and took himself off for a wee. As he negotiated his way back to the sound of crunching and munching from his sister, I tossed him a handful of conciliatory nuts to keep him occupied while he waited.

## The pig in the labyrinth

Professor Mike Mendl responds to my story like a seasoned parent.

'Initially, your pig might well have been screaming to express hunger,' he says. 'But if you're rewarding that behaviour they will learn from it.'

'I didn't feel I had much choice,' I tell him.

'If it was a child you would ignore it.'

I know he's quite right, of course. Maybe Roxi would've desisted had I not given in and served breakfast under my own terms. But then I am quite sure many households within a 500-metre radius would've countered by serving me with a noise abatement notice. Regardless of my handling, I am interested in the fact that each pig sought to manipulate the situation to their advantage. Did that make them smart, sneaky or both? As the Professor is one of the country's foremost experts in pig cognition, he seems pleased to move on from my questionable swine-herding skills and on to his specialist subject.

'The question of whether some animals can be deceptive began with a study of chimps,' he says. 'The original study featured a chimp called Bella. The researchers placed food in a certain place in a field for her. She would take the food and then return to her group. Eventually, the adult male sussed her out, followed her and took the food for himself. Next time, Bella then showed an apparent deception by leading him away from the food before rushing back to get it.'

It's a story that's as cute as it is enlightening, but Professor Mendl is keen to point out that this doesn't mean chimps could mask a winning poker hand. 'It's sophisticated,' he says, 'but we're not certain that what they're doing is intentional deception. It's just because they're primates and they look a bit like us that people are ready to draw that conclusion. With pigs,' he suggests, 'we are more sceptical.'

In his research, and careful not to fall into the trap of *wanting* to believe that pigs process thoughts and feelings just as we do, the Professor and two colleagues set up a maze with a food source hidden in a one location. Releasing a pig into the maze, they observed it forage around and figure out how to find the food. On the second visit, the pig demonstrated a sharp sense of spatial awareness as much as a memory by heading straight for the source.

For the next stage of the task, a bigger, more dominant companion followed the informed pig into the maze.

'Over trials, the bigger pig twigged that the other one knew where to go,' says the Professor. 'Eventually, when the informed pig went to the food, the bigger pig followed and displaced it.'

I nod, mindful of the way that Roxi displaced Butch from the breakfast bowl, effectively an all-out assault.

'After that happened a couple of times,' the Professor continues, 'the one with the knowledge would not go to the food bucket straight away. Now, one possibility is that the informed pig thought, "Ah, the dominant pig keeps getting to the food and so I'm going to do something different." On the other hand,' he says, 'the informed pig may have just been avoiding the dominant pig because negative things kept happening. Then, once the dominant pig was out of the way, it hurried back for the food. Either way, it's still a knowledge thing. They're picking up what to do by association. Once they understand what predicts whether they get – or fail to get – the reward they can be very quick to modify their behaviour.'

I consider my experience in the light of Professor Mendl's findings. Did Butch and Roxi deceive and exploit each other to get a first crack at the breakfast bowl? In my view, each one had processed the situation they were faced with and worked out how to put themselves first.

According to the Professor's findings, the key to understanding what makes a pig tick is to recognise its ability to *learn*. He tells me, for example, how a colleague found some evidence that pigs can grow to understand the concept of reflections. This involved releasing a pig into an arena with a mirror placed just beyond the far end of a barrier. From a certain angle, it enabled the pig to see a food source on the other side. Rather than crashing into the glass, the Professor tells me, the pig appeared to work out how to use the reflection to guide it back around the far end of the barrier in order to reach the food. Whether a pig can recognise its *own* reflection, which would suggest a degree of self-awareness, we simply don't know, but we both suspect there is a great deal going on between the ears.

Professor Mendl and his colleagues continue to devise fascinating ways to investigate what degree pigs can be said to be smart or sly. To the best of my knowledge, and under deeply unscientific conditions, all I can say is that I knew two that had repeatedly taken advantage of me.

## Wendy's world

'I do think pigs are very knowing, but there is a big variation between smart pigs and thick pigs. It's the same with people, really.'

Wendy Scudamore is so passionate about pigs that it guides her outlook on life. Hidden away on a bucolic farm on the slopes of the Golden Valley in Gloucestershire, her cottage overlooks steep-sided hills and pockets of forest veiled in early-morning mist. Wales is just one field away to the West, with a view of the Black Mountains towards Brecon and a vast, ever-changing sky overhead. On a visit one morning in late spring, I am stopped at the gate by an advance guard of little piglets. They're rooting around on the farm track for what's left of a scattering of feed pellets. They're so locked into their search that I can't be sure if they're aware of my presence. I suspect they probably are.

Five minutes later, having entered on their terms, I knock at the farmhouse door to be greeted by a dark-haired, elegant figure in muddy overalls patched at the seat with silver duct tape. Wendy has lived here since 1992, but it's more than just a home. She introduces me to her son, just back from university and off to walk his dog, while out in the yard and across the fields and paddocks are the pigs that make this a remarkable little world. As she puts on the kettle for tea, checking I'm OK with fresh goat's milk as that's all she has, I am struck by how so many of her family pictures feature children through the years, cuddling piglets or being photo-bombed by lumbering fat sows. Wendy is, without a doubt, a pig person, and I am here to be enlightened by her.

'I used to promote the intelligence of pigs by taking an agility course around agricultural shows,' she tells me over a distinct but enjoyable cup of tea. 'I had one lovely pig who used to do it to music. She would follow me round and I just sort of told her what to do. I wanted to show that they aren't just lumps of meat you can stick in a pen, rear and eat. A pig is a sentient, emotional and very affectionate creature, and I hoped that it would encourage people to become more concerned about the pork that they buy.'

As the owner of an unruly Miniature Dachshund and a selectively deaf Greek rescue, I am heartened to learn Wendy believes that, like dogs, some pigs are more amenable to picking up tricks than others.

'In 2010, I was invited to train three little ginger pigs to appear at the Cannes Film Festival,' she tells me. 'I did it with a clicker, which drove the soundman mad, but one pig in particular would do everything I asked. Brad was fantastic. He would sit and wait for me to tell him what to do, whereas the other two just wouldn't listen. Nicole Pigman was the worst,' she says, and I try to keep a poker face. 'I just couldn't get her attention. They were from the same litter, just different genders.'

'Is it a boy-girl thing?' I ask.

'The third pig was a boar, and though he was quite smart, it was Brad who stood out as the star. I think it came to down to concentration span,' she suggests, and then tells me Brad is still alive and well and enjoying his autumn years up in one of the paddocks. She talks about him like an old thespian friend in retirement. As her stories continue, it strikes me that Wendy has formed a lifelong bond with every one of her pigs that begins with her recognition that these are creatures of significant intelligence.

## After Bertie

With my limited success in dog training, I know that treats are a key motivator. The clicker is only effective once the dog associates the sound with something that makes it drool, but do pigs operate on the same basis? When I ask Professor Mendl, I am surprised and not a little delighted by his considered view.

'Pigs are motivated in my tests by the food reward,' he says, 'but the experience of the test itself is also rewarding. We don't know for certain, but pigs seem to enjoy it. When we work with them in the maze test you get the impression they are keen to do the task and not necessarily just for the food. It's difficult to disentangle,' he continues, 'but when we work with them for several days, they learn in what order they're supposed to leave their pens and start to queue up accordingly. So, one will think, "Well, Bertie is first and then it's me." They learn that sequence and know when to come out.' The Professor tells me that he has even seen cases where one pig will push another out of the way if they're in the wrong order.

'So, they're switched on and also determined,' I say.

'It suggests they're motivated by something to do that is reasonably interesting,' the Professor replies, choosing his words with the precision his work demands.

At the same time, I think of Wendy and her agility pig in a show ring of straw bales, and wonder who had the most fun.

## Egg heads

For a while, Butch and Roxi spent their days cohabiting happily with my chickens. The pigs proved to be a fantastic fox deterrent, which made this set-up seem ideal. With their sleeping quarters at one end of the enclosure, and the coop at the other, the two sides quietly learned to get along. The pigs had a size advantage, of course, and so if they moved into a spot to root around, the chickens would duly dance out of the way. Nevertheless, as rescues from a battery farm the chickens didn't suffer fools. They could get cross very quickly, proving sharp with their beaks, and so a mutual respect evolved.

Until, that is, the pigs cottoned on to the fact that the hens laid treasure inside their coop.

I had considered that the eggs might be an issue. Then again, the nesting box was tucked at the back of the hen house. I didn't think Butch and Roxi would be wise to the reason why the hens took themselves in there each morning. Through my eyes, the system was pig-proof.

In some ways, the chickens only had themselves to blame. Among the flock was a vocal leghorn who liked to announce that she had laid successfully by squawking for several minutes afterwards. Maybe there was something in the tone that eventually told Butch and Roxi that it was worth investigating, which they carried out much like a police raid on a drug den at dawn.

As Roxi effectively rammed her great head inside the coop to snatch an egg, followed by her colleague, the chickens exploded in feathers and anger. Both she and Butch endured a flurry of pecks as they retreated to their quarters with yolk dripping provocatively from the corner of their mouths. With a heavy heart, as I watched from the bedroom window, the chickens formed a vigilante mob at their door, before wisely retreating, one by one.

In a bid to help them defend their eggs, which was effectively my rent for their tenancy, I extended the legs on the coop. There was no way that Butch or Roxi could negotiate the narrow ramp that led to it, and indeed I was correct. They just worked on undermining the legs until the whole coop tipped over.

For a week, I tried different means of preventing the pigs from gaining access to the eggs. Planting the feet deep in the ground just lowered the roost enough to give them access once again, while Butch discovered that headbutting the underside could sometimes cause the eggs to jump and roll out. Every time, driven by their prize, Butch and Roxi found a way. It became a daily mission for them, often requiring ingenuity as much as physical force, and neither would give up until they left with egg on their faces, as intended. Eventually, before someone lost an eye or the pigs evolved their taste from egg to chicken, I evacuated the hens and placed the coop on the lawn. Peace quickly returned to what was left of my garden, along with the recognition that for all my efforts to deter the sly and ingenious fox there was nothing that could beat the tenacity of a pig.

## The journey and the destination

It sometimes seems to me that the domestic pig has it all figured out. Everything they do in between waking and flopping down together at dusk is undertaken with such enthusiasm that any treat at the end seems almost secondary as a reward. Watch any pig at work in the soil and you could find yourself observing for hours on end as they excavate relentlessly. Using their snout as a shovel, that pig will till, chisel and sweep, and send frequent sprays of soil up into the air. You'd be forgiven for thinking the scale of the undertaking must mean they've locked onto something pretty special. Towards the end of the dig, having formed a crater, you might just see the pig's hindquarters and curly tail as it mines its way towards the prize ... invariably a half-rotten acorn, gone in a crunch. Was it really worth the effort?

'They're just very investigative,' says Professor Mendl, whose accounts so far of the tasks he sets for pigs not only provides me with enlightening data but sounds like enormous fun for all involved. 'We're always putting things into the pens, changing or modifying aspects, and they'll always find ways to remove or destroy them. Once, we had a drainage cover in a concrete pen. It was flush on the floor, and there was no apparent way they could lift it. But every day, I would go in and it would be gone. I'd have to look everywhere, and eventually find it tucked behind a panel or buried under straw.'

I express a distinct lack of surprise at the conduct of the pig in this respect, but I am interested to know what drives such investigation.

'Probably food,' he suggests, 'but they are also information gathering about lots of other things, and there's something rewarding for them about this behaviour. Pigs can pick up on olfactory signals in the ground about other pigs, or where good foraging can be found, and all this comes through while they're searching around.'

'Or hiding drain covers.'

'We read it as mischief-making,' says the Professor, 'But I don't think they're intentionally trying to wind us up.'

I take his point, and I'm well aware that in their enthusiastic pursuits Butch and Roxi never set out to give me a headache. It's perhaps just that it always fell to me to deal with the damage, and it never seemed to end.

'Pigs are also very persistent,' he offers. 'A lot of species can be like this, but the nature of their food requires quite a bit of digging. Pigs just demonstrate this in a clear and obvious way.'

Listening to him, I realise that if I want to understand a pig then I need to stop comparing their behaviour to humans. Toppling a chicken coop or stashing a drain cover might cause us to sigh, but for the perpetrator, everything is there to be examined. What is a bombsite through my eyes is really a forensically combed food scene. The excavation has been conducted with huge enthusiasm, no stone or clump left unturned. It might look like a mess to me, but to the pig it's the sign of a job well done. And even if such labouring yields nothing more than a withered weed root, well, that's the icing on the cake. After the bliss of the dig, it must taste like heaven.

## Immovable objects

On their greatest escape, it took the best part of a day to locate our missing pigs. The trail of destruction from the splintered gap in the fence onwards had gone cold midway down the lane into the village. Together with my wife and children, I combed the fields and copses in search of Butch and Roxi. We called their names repeatedly, which did nothing more than draw people from their houses to join in with the search.

By the time we found them, in ancient bluebell woodland flanking the far side of the village, our search party was almost 50 strong.

Both pigs seemed unaware of the fact that they were busy rooting up a protected species of flower. It wasn't lost on me, however, and so it was with some urgency that I endeavoured to steer them out of trouble. Neither Butch nor Roxi would budge, however. Such was their focus on the dig that they didn't even seem to register my presence. As word spread of the discovery, they remained oblivious to the growing number of search volunteers who picked their way into the clearing.

'Time to go,' I told Roxi, appealing to her directly by crouching in front of her. The pig responded by prising up a batch of bluebells.

Had this been a dog, I could have grasped it gently by the scruff and steered it out of trouble. A pig provides no such purchase, and it wasn't just me who tried to push and pull her into line. Roxi simply transformed herself into a living anchor. One that snorted angrily when she'd had enough hands-on attention.

Butch proved a little more compliant. Being of a more timid nature, he bowed to my begging while a couple of villagers eased him off the flowers, only to turn around at the first opportunity and return to the work at hand. Someone handed me a dog collar and lead, and when the collar didn't even come close to reaching around their great necks I honestly thought we might just have to leave them to it. I would pay the fines and the pair would go feral and become part of village folklore. Bodmin had its fearsome beast. We would have two wilful pigs with a taste for rare flora.

Eventually, a villager who kept pigs of his own showed up with a wooden board and a bucket containing a handful of feed. He handed the bucket to one of my daughters and instructed her to stand with it in front of Roxi. The pig paid her no attention at first. Then our good Samaritan placed the board to one side of Roxi's field of vision and tapped her lightly on the flank with the stick. With a grunt, the pig lifted her snout from the soil and sniffed the air. With the board blinkering her field of vision, and her senses locked onto the smell of pig feed, she took one step forward, followed by another, as my daughter with the bucket, the villager in charge, and the crowd behind Butch began the slow procession home.

When I recount the story to Professor Mendl, I expect him to confirm the age-old belief that pigs are intrinsically stubborn. Instead, he provides an explanation that seems wholly reasonable. It also reminds me that had our pigs been genuinely mini as advertised, things would have been more manageable.

'It's very difficult to persuade a pig to do something,' he agrees. 'But that's because of their size. If they want to be stubborn then you know about it, but a cat can do exactly the same thing and you just pick it up. The fact is a lot of animals will stand their ground or ignore you,' he adds, and places cattle squarely in that frame. 'It's just with the bigger ones that stubbornness is magnified.'

## Be more pig

From an early age, we call upon the animal world to help us make sense of human behaviour. From the untrustworthy snake to the wise owl, the wily fox and grumpy bear, we tell stories featuring such characters to help define our identities and values. Often drawn from myth and fable, such tales are entertaining and serve a useful purpose, but does it distort our relationship with the animals themselves? My Dachshund can be loyal to order, if I have a treat in hand, but at any other time he'll switch allegiance for the nearest available lap. The same goes for the supposedly brave lion, stubborn donkey and a whole host of creatures that have been categorised by an attribute that doesn't necessarily reflect their true nature.

Like Wendy Scudamore, I have come to see pigs as individual characters. Some strike a charming first impression, others are bold, belligerent, playful or free-spirited, but spend a long period of time with any pig and their personality will take shape. It's as elaborate as ours, but with a mind that operates on a sensory level in which smell opens up a world we cannot comprehend.

So, let's not dismiss the pig as greedy or messy, when the reality is this hard-working creature has developed a passion for turning the earth in search of a mere scrap of food. Can we really call a pig lazy when they're prepared to dig all day on detecting something tasty in the soil? I might've marked my two down as wilfully destructive, but in their minds laying waste to the lawn is a measure of progress and a means of mapping their environment. Yes, they can be obstinate, but then again so can I, and if I shared their ability to focus on a task and not give up until it's done, I'd be an awful lot more productive.

In many ways, we have done a great disservice to the pig by associating it with so many disparaging adjectives. The animal is far from stupid, ugly or ignorant, though whether any are sexist we can only guess. As we'll see in the next chapter, when it comes to matters of the heart they certainly view life differently.

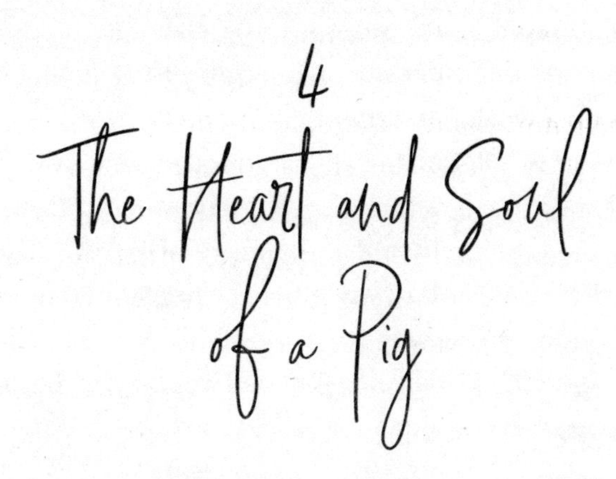


## Holly and Poddy

Over tea in Wendy's farm kitchen, with her three Collie dogs lounging at the foot of our stools, she tells me a story about two pot-bellied pigs, which stays with me all day.

'Holly and Poddy were the first pigs I had,' she begins. 'They lived at the bottom of my garden, and then I moved them out to the field. They were quite nervous, as pot bellies often are, and always stuck together. Holly was slightly the underdog. She followed Poddy everywhere for the first six months of their lives.

'And then suddenly Holly died,' she tells me so abruptly that I catch my breath. 'I found her on the stable floor. Poddy was with her, and she was heartbroken. Now, whenever a pig passes away, I always leave them with their friends for 24 hours. Even the dogs will go in. That way, everyone becomes aware that they've gone, and there's nothing there. The whole time, Poddy lay with Holly,' says Wendy, 'and when I took Holly away, she literally cried.'

Wendy holds her cup with both hands as she tells me this. She sips her tea as if to take stock of her thoughts.

'Poddy just whined and didn't want to come out,' she continues. 'When she did, she would walk around the yard, go back in and start whining again, and this went on and on. I had other pigs, and Poddy integrated with them, but who knows if she ever got over it. You can't ask her. I don't know if she ever really made another friend.'

In a bid to help her grieving pot belly, Wendy tells me that she put her 'in pig', which is breeder speak for pregnancy. 'I thought that if Poddy had a litter then I could keep one back and she would become just as attached as she was to Holly, but it didn't work out. She went to full term, her milk came in, but then she fell poorly. The vet came out. He examined her, but found nothing there.'

Wendy pauses, and picks up on my puzzled expression.

'If something goes wrong early during a pig's pregnancy,' she explains, 'the mother can reabsorb the dead foetuses in her litter. In this case, Poddy lost everything, but her milk still came through. She was sick for a long time afterwards, and though she recovered eventually and lived until she was 12 she wasn't the same without Holly. She would snuggle up with other pigs, but I could always separate her easily. She just never had a friend like Holly again.'

Wendy is careful not to humanise her pigs. When she says they cry, she always stresses that there are no tears and it's simply a description of the sound. In listening to her story, it's easy to relate to Poddy's lifelong plight. 'Some pigs are more emotional than others,' she adds, 'just as some are more sociable. But I have never seen the level of pain that Poddy displayed.'

In addressing the subject of a pig's emotional range, I am well aware that we are in a realm of both science and wonder. Professor Mendl rightly deals in data. He qualifies his responses on this subject by saying we don't really know for sure, and that's just as it should be. However, having seen Roxi buck with apparent glee whenever she slipped the gate into the garden, or watched Butch nuzzle my son when he was very small, I'm quite prepared to accept that pigs have feelings, too. They might not have the same understanding of joy or fear, but we'll always use words to describe what we're seeing based on our experience. Nevertheless, when one of Wendy's pigs dies, I can understand why she allows time for the other pigs to register the loss. She might be drawing on our bereavement process, but I'm quite sure we're not alone in drawing comfort from it.

If anything, by accepting that pigs are emotional creatures just like us, the bond between us becomes closer. One way to explore this is to consider their model of family. It's very different from our own, and even alien in some ways, but tells us a great deal about pigs as much as about ourselves.

## The matrilineal group

Welcome to a world in which the sows stick together. For the domestic pig living freely, just like its wild boar ancestor, a family unit is made up of two to five females and their offspring, along with one boar. The young sows tend to stay with their mothers, while on maturity, the boys instinctively seek new pastures. Collectively, Professor Mendl tells me, this set-up is often known as a 'sounder'. He goes on to refer to such a group using an altogether different term. Still, it's one that helps me process what's going on.

'The adult male is dominant,' he begins, 'but his role is to protect his harem during the breeding season. Outside of that, he may drift off a bit.'

I want to know if the boar steps back simply because the sap stops rising. The Professor tells me that size can also be a factor, in that he might gather too many sows under his wing than he's able to cope with and then leave himself open to a challenge.

'When young boars depart from a harem they emigrate in small bachelor groups,' he explains, and such phrasing is not lost on him either. 'One might be successful in taking over another group, or hiving off sows from a large group to create a harem of his own.'

'Is it a peaceful takeover?' I ask.

A wry smile from the Professor leads into his answer: 'During the breeding season, a boar will want to retain access to his females. So if another male comes into the group during this time they are strongly repulsed. They will fight, but first, they do some displaying, which often takes the form of parallel walking. It's a way of showing off to each other, much like red deer, to resolve any differences. If that doesn't sort it out then things will escalate into head-to-head fighting and biting. Eventually, one will win, which really means the other decides to leave.'

Out of season, he tells me, a group takeover might well occur simply because the boar in charge has taken his eye off the ball. He describes how another boar might come in, draw away some of the females and form a group of his own. 'Within this structure, there's also no need for the sows to go looking for a male. If the group becomes too big for one boar, another male will come in and take some away. How they persuade them is a good question, but often it's the younger members who go, and a natural equilibrium is achieved.'

As a social model, the Professor tells me, the matrilineal group minimises inbreeding by encouraging young boars to form sounders of their own, while allowing the pig in the wild to flourish. In effect, it's a tight family unit, often comprising mothers and their daughters, and a leader primed to lose it all when he ages, over-reaches or becomes complacent.

I can't help but picture such a structure from a human perspective, and smile to myself at the idea of the alpha boar who is kept in check this way. I can also see how the arrangement perfectly suits an animal that depends on close social interaction with its own kind. It's as stable for the sows as it is fluid for the boars, self-regulating in size and organised in such a way as to promote security and lasting companionship.

## In the wild and on the farm

If pigs are hard-wired to form small groups in the wild, what happens when they're raised domestically? This is a significant field of interest for Professor Mendl, who carries out his behavioural research with the intention of improving pig welfare in captivity.

'Pigs have an innate tendency to repulse pigs they're unfamiliar with. So, you often see conflict on farms where pigs are mixed without warning. Olfactory cues will trigger it,' he says, and once again stresses the central role that smell plays in this world. He also explains that adult boars are always carefully controlled in a domestic environment, and that only growing pigs that have yet to reach sexual maturity are put together in this way. Nevertheless, when anything from 40 to 60 pigs find themselves sharing the same space, it's inevitable that these animals, with much smaller grouping needs, will undergo some kind of social shakedown.

'They might not react to begin with,' the Professor explains, 'but then one will pick up on the presence of a pig that smells unfamiliar. That's when the skirmishing starts, and it can be between males and females. In reality, they're just directing aggression towards something that smells different, but a period of fighting can develop that lasts for 30 or 40 minutes, and it can be chaotic.'

'There is data to show that when a fight kicks off, a third pig from the group might intervene to show support,' he adds, which makes me think of Friday-night tussles as the pubs spill out, but I also appreciate what it says about a pig's sense of identity as a collective. 'It finally settles down when they're exhausted, and at that point you'll see the new group coalescing.'

Professor Mendl then describes a post-fight phase that strikes me as rather sweet. 'They tend to lie down together and become familiar with one another, using cues like smell and vision,' he says. 'With the threat gone, the new pig is absorbed into the group.'

## Mix, match and move on

The fields and paddocks surrounding Wendy's cottage are dotted with dwelling places for her pigs. Some are built from timber, others from quarry stone and look like they have stood for a century. Then there are the outbuildings, all of them carpeted with straw and home to individual huddles of pigs. While each group has masses of space, and some are free to roam, Wendy still manages the pig population to keep the peace and allow them to live their lives to the fullest. She has a light touch, and carefully considers what pigs are suited to each other in terms of their individual qualities.

'Even if pigs don't get on during the day,' she says on the subject of her approach to integration, 'they'll eventually sleep together at night. So if I have a new pig, I'll make a place for it to sleep outside until the others let it into the bedroom.' She adds that assimilating pigs on neutral ground is the most effective way to minimise conflict. 'There will always be a little bit of fisticuffs and then it's fine. But if you put a pig into another pig's ground, male or female, they will fight for it. They are very territorial.'

Even as she tells me this, I note for myself that Wendy has a considered eye. Through the window I can see two pigs happily investigating a grass slope beside an outbuilding as if they've been tasked as a pair. Does she ever have problems where one pig turns on another?

'Oh, some can be bolshy and pick on everyone,' she says, 'but they're not bad-tempered. I just interpret it as them being top dog and full of themselves. They certainly become more dominant as they mature, but then they get dominated in old age. That's when I have to keep them with more appropriate pigs,' she adds, and tells me that her elderly stunt pig, Brad, no longer rules the roost but lives happily with a 'castrated little kunekune'.

## Protection racket

From Wendy's kitchen, beyond the courtyard gate, I can see pigs in small groups. They occupy former stables fronted by farm gates. Piglets scamper about. Some are small enough to fit through the bars and frolic. If there is a system in place, it's relaxed around the edges. I take this as a reflection of the pigs' capacity to self-organise as much as Wendy's trust in the community as a whole.

'They're very supportive to one another in a crisis,' she points out when I observe how close-knit they seem. 'If I need to inject piglets [against disease], I shove the mother out first. Otherwise, as soon as I start picking up the piglets they'll start squealing and then all the sows will come to the gate. They will not have you hurting or stressing the young. Even if the piglets don't belong to them, their maternal instinct is very strong. Take the wild boar out in the forests here,' she says, gesturing towards the window. 'Dog walkers often find themselves under attack because a boar has become alarmed at their presence and all the others come to their rescue.'

## The reporter in the pigpen

In some ways, I decided to write a memoir about my experience in raising two supposedly small pigs as a form of therapy. It took me the better part of a year to complete the book, and make sense of the ongoing chaos that dominated our day-to-day lives. During that time, Butch and Roxi went from the size of a pair of shoes to a pair of Labradors. Nine months later, on publication, they had timbered up so significantly that when my editor came to visit he stopped dead in his tracks in front of their enclosure.

'They're massive,' he said, as if it might have escaped my attention.

Relatively speaking, Butch and Roxi were just standard mixed-breed pigs. Still, when the book came out their size became a story. Several newspaper photographers took their picture. Butch and Roxi accepted these visits with good grace. If they were dozing, they'd emerge from their sleeping quarters to greet the visitor. Roxi would be first to approach, grunting and sniffing. Once satisfied, or distracted, by a scattering of pig nuts, the photographer would be free to get the required shot. One even got down in the mud in composing his shot in order to make them look even heftier.

Maybe that was a turning point for the pigs, because when the next visitor arrived – a television news reporter – his piece to camera didn't quite go to plan.

Having listened to Professor Mendl stress how much pigs are informed by smell, I can only think the reporter carried with him the scent of a pet on his suit trousers – or perhaps a sense of entitlement that rubbed them up the wrong way. For he followed me through the gate, showing no sign of trepidation, but also failing to pay the pigs any attention. Instead, turning his back to them, he faced the camera operator, who had opted to stay on the other side of the fence. At the time, staying out of the enclosure myself, I did wonder if it had been wise to wear such smart attire. In my experience it was impossible to step into their enclosure without getting dirty to some degree. Still, he looked assured on asking the audience to picture the pigs behind him as tiny, toy-like creatures.

Even before Roxi launched herself at the reporter, knocking him off his feet with an angry bellow, I knew that she was upset. It was an ember in her eye. Just a glint as she swept the ground with her snout and then mounted her attack like an incandescent bull. The poor man practically turned a cartwheel in the air. In horror, I watched his phone and various memory cards slip from his pockets and into the slop, followed by the man himself. I had never witnessed this level of aggression from Roxi before. She could be bad-tempered and grumpy when hungry. This was very different, however, and I scrambled over the fence fearing she might weigh in on the guy as he floundered in the mud, trying to get clear.

But instead of seizing the advantage, Roxi simply stood her ground for a moment. Then, as if satisfied that she had made her feelings known, she metaphorically wound in her neck and rejoined Butch in scouting the ground for stray feed pellets.

'Are you OK?' I asked uselessly, and offered him my hand.

Ever the professional, despite being striped in slurry, the reporter clambered to his feet and duly picked up his belongings. Minutes later, standing on the other side of the fence, he recorded a short, less enthusiastic piece to camera while I held his jacket off-camera and wondered if I should offer to dry clean it. The item went out the next day on the evening news. It sparked a few more requests from media agencies. I declined them all. This was Butch and Roxi's home. They didn't necessarily live a quiet life, but the episode showed me that pigs are incredibly sensitive to their space. Ultimately, they need to feel that anyone who shares it can be trusted.

## Social security

'Nobody lives alone here,' says Wendy, when our conversation turns to the instinctive need of a pig to be among its own kind. 'They're social creatures. They're cuddly, they want to be beside other pigs. They need to be able to see and feel and be up against another body. They're not loners.'

In the pig world, I first came across Wendy in her role as a voice of truth about minipigs. At a time when the internet was awash with what was effectively cute pictures of standard piglets, she took to task those who were selling a dream of a pocket-sized breed, often as a single pig, without apparent consideration for the long-term consequences. Even now, she is forthright on the subject and does so in the name of animal welfare. It isn't just a case of raising awareness about their potential size, she stresses. Regardless of how big they might become, a pig needs both space *and* company.

'They're not like dogs,' she says. 'If they start off being taken from the litter to live alone, they can't communicate and have to live like something else that isn't them. That's so wrong, it goes against their instinct.

'People tell themselves the pig is happy in that situation,' she adds, and cites cases where minipigs outgrow apartments but have nowhere else to go. 'In reality, they've just adapted to an alien situation.'

# The complete pig

When Butch and Roxi outgrew our expectations, we were fortunate enough to have the space to accommodate their needs. Their welfare was our priority, even though we hadn't banked on keeping fully-grown pigs in our back garden. While our research had been lacking, we had at least registered that pigs valued company of their own kind. While it meant double the destruction, our pair proved inseparable. In slumbering together, Butch would arrange himself so that his snout was wedged between Roxi's flanks. It was a nose-to-tail arrangement that suited them both, despite her noisy flatulence, which I think he secretly found to be a comfort.

As for their waking hours, once the battle for breakfast was over, both pigs would settle into a routine in which

they did everything together. From foraging in the soil to snoozing in the sunshine or breaking through the fence, Butch and Roxi relied on one another for companionship and security.

In a sense, they complemented each other. Being the larger of the two, Roxi was certainly more assertive. Then again, Butch possessed a greater patience. In autumn, when the apples fell from my little tree, Roxi would be quick to become hysterical if one dropped on the wrong side of the garden fence. Butch, meantime, would simply wait for one of us to investigate the source of the noise, knowing that the apple would be retrieved and split fairly.

It was in digging that their contrasting talents really came into their own. Where Roxi was skilled in brute force, with a skull that could shovel away soil in great heaps, Butch possessed a more sensitive snout. He would often take over on the last leg of an operation to excavate to new strata of surprises, and then stand back as Roxi took first pickings. They worked very well together. I liked to think it was by design, and that the two had arrived in this world dependent on each other. In truth, pigs are smart creatures that read each other closely through certain senses far sharper than ours, and then figure out how to make a whole greater than the sum of their parts.

## Meet the Swedes

If pigs use their differing personal qualities to lock together as a group, what influence does their breed bring to bear? The fact is most breeds are selected for the quality of their meat, and behavioural traits are often overlooked. When I ask Wendy for her view, she suggests it's time to meet her pigs. As we slip on our boots, with the dogs turning circles of excitement around us, she tells me we should start with the top field.

'I have some pigs that I imported from Sweden,' she says on picking our way up the farm track.

From her tone, this sounds like something she might not do again.

'What are they like?'

'Well, it might be me projecting, but they are completely different in character to my other pigs.' Wendy takes the lead as she tells me this. We pass a young man with sea-blue eyes sitting on the verge. He's wearing overalls and a flat cap fringed by ginger curls, and nods pleasantly at Wendy. I'm guessing he has something to do with the working farm that seems to operate in fields and pastures adjoining her land. Either way, he overhears her mentioning the Swedes and seems quietly amused.

'I thought they would be nice for pig enthusiasts,' she continues, and explains that she first brought them into the country with a view to breeding them. 'But I didn't sell any.'

'Why not?' I ask.

Wendy glances over her shoulder, just as we reach the crest of the path. 'Because they're *appalling*!' she says, and invites me to join her at the gate overlooking a gently sloping apple orchard.

I pick my way through a puddle and lean on the upper rail. The rainwater has run off the field and pooled in a basin in front of the gate. It forms a kind of moat, which I'm not sorry about when I spot half a dozen creatures with matted, yak-like coats picking up on our presence. They look like little pigs in weird, hairy fat suits, but most striking of all is the fact that there's something off-message about their manner. None come trotting down to investigate, as I might expect from any pig. These Scandinavian sows just cease their inquisition of the earth and simply stare as if we're here without first seeking their permission.

'They don't want affection,' Wendy tells me, and wearily climbs the gate. 'They're not interested in people, they just think they're superior. And they'll take you out if you try to go near them when they have piglets,' she adds before gesturing for me to follow.

I hesitate for a moment. Wendy is already sloshing through the muddy soup towards higher ground.

'Do they have piglets?' I ask after her, but I think she chooses not to hear me.

By now, the Swedish pigs have started honking. I realise I am investing human qualities in the noise, but they sound totally affronted. Feeling like I should have brought a gift or something, I follow Wendy over the gate and into their field. As I negotiate the water, she calls out, 'Come, *PIG*!' in a shrill voice. In response, the honking becomes more animated and several of these squat and dark wretches begin to gravitate towards her. They don't exactly hurry, however – as Butch and Roxi might have done on the assumption that I had something for them to eat – and indeed pull up at some distance from us.

'They seem wary,' I say, struck at the same time by their prominently upturned snouts and glimpse of tombstone teeth underneath. 'And kind of visually challenging,' I add under my breath.

'The girls are very timid,' says Wendy, and begins to name them in turn. 'That's Malm. Over there is Ektorp …'

I glance at Wendy side-on. The names sound familiar. In fact, I'm pretty sure that one of the pigs shares a name with my sofa.

'Ikea?' I ask.

Wendy nods as if she had no other choice.

'Lowie is the only one who's sure to come and see me,' she says, looking around, and then her face lights up. 'Here he comes …'

I follow Wendy's line of sight. Across the orchard, huffing and puffing as he negotiates a terrain as crumpled as a bedsheet, a shaggy off-white boar approaches us. He's only slightly larger than the girls, which isn't saying much, and looks like he's wearing a lion costume tailor-made for him.

'Lowie has small-man syndrome,' says Wendy, and calls the pig one more time. 'I handled him a lot when he was young, but then he got boarish and tried to dominate me. He just wasn't afraid at all.'

By now, Lowie has begun to make a weird clashing sound. He's moving his jaws, I realise, for the tusks so characteristic of his gender are bobbing up and down and I can see a cluster of front teeth.

'Is he chomping?' I ask.

'And marking his territory,' says Wendy as the boar rubs himself against a tree. 'Protecting his girls, though they're all spayed so he can't cause trouble.'

The sows, meanwhile, are gawping at us in undisguised antipathy. Having watched us spoil their party, it's as if they're waiting for Lowie to take us to task. While I stay quite still, just in case the boar showcases a sudden temper, Wendy continues to coo at him with one hand outstretched. Slowly but surely, to what sounds like knives sharpening inside his mouth, Lowie gravitates towards her. He's beginning to froth at the corners of his jaw, I realise, but Wendy assures me he's just chomping on his molars to sound intimidating. At her feet now, the little pig peers up at Wendy. All of a sudden, I see a look in his eye like a little boy in need of some attention, which is exactly what she delivers in the form of a scratch behind the ears.

'I can probably give him a tummy rub,' says Wendy as the sound of grinding teeth and huffing turns to squeaks of joy, 'but they are just so suspicious.'

I glance at the girls. I'm not sure trust is the issue here. They just look as if they're witnessing a scene of wholesale treachery by the boar whose one job it is to protect them. Unlike any pigs I've come across before, they keep a distance between us as if any kind of contact could prove ruinous. Wendy tells me that they don't show such disdain for other pigs, which further marks them out as different from any other breed she's known.

'Can you mix them? I ask.

Wendy chuckles, which is enough to give Lowie a start.

'I had to remove the Swedes from the English pigs,' she says. 'They were stroppy and extremely sexually orientated. In season, the Swedish girls would try to mount the British sows. They put up with it, but I could see that trouble was in store. So now they're up here, and seem much happier among their own kind.' Wendy draws my attention to the fact that the Scandinavian sows sport hairless stripes along their upper flanks. 'That's where the girls have been riding each other,' she says. 'They've worn away the bristles.'

'Oh,' I say, and now that's all I can see.

'My conscience wouldn't allow me to sell the Swedes,' she adds. 'You can't offer pigs like this to people!'

Watching Wendy indulge an inherently cautious little boar, as his harem look on in utter disbelief, I wonder if secretly, she's content to keep them all to herself. The Swedes are certainly a breed unto themselves, but despite the challenges they present there is a bond between them that's as unique as it is unbreakable. Even when we turn for the gate, leaving the girls to glower at Lowie, it's clear they'll quickly forgive him in the name of group harmony.

For pigs of any kind, no matter what their number, gender, personality or the characteristics inherent to their breed, there is nothing more important than family. This, in effect, forms the collective heart and soul of their lives, and allows the character of each and every pig to shine.

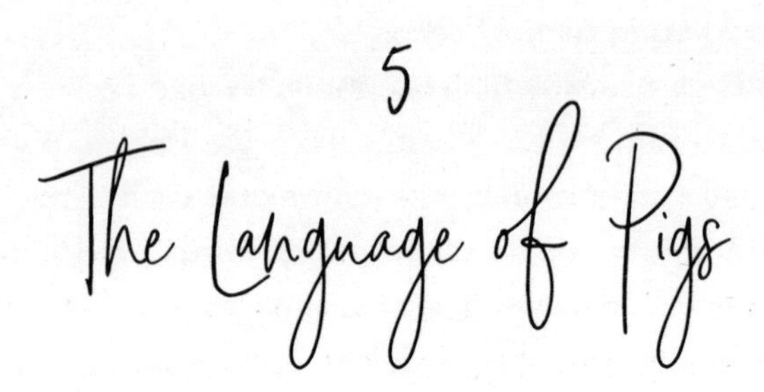

# 5 The Language of Pigs

## Rocky

'He was the first piglet that I ever kept from a litter,' says Wendy as we amble down from the orchard. 'Rocky had free run of the yard, and never ever got shut in. He was huge – a whopper, but a darling – and belonged to my young son. If something had happened at school that upset him, he would come home and talk to Rocky. They were very close. And then sometimes Rocky would talk to me.'

She points out a nearby track and tells me that the pig would often take himself there to bask in the sun. 'If I spoke to him, he would immediately answer. I'd say, "Oh, Rocky," and he would honk, and we would have a conversation. I could lie down beside him and have a cuddle and he would grunt to me.'

'Was he talking to you?' I ask.

'Oh, definitely,' she says, with such a note of sadness that it's clear he has moved on. 'I got terribly attached to him.'

## Conversations with a pig

Wendy talks about each and every one of her pigs as a personal friend. Even so, she is quite realistic in recognising that they speak a very different language. If she humanises her pigs in any way, she's quick to spell that out, which I read as a mark of healthy respect for an animal she cherishes dearly. Nevertheless, the connection she describes through verbal contact is clearly one that brings pig and keeper closer together.

'You can invest what you want in the grunt,' she says, well aware that her conversation with Rocky was based on sounds and not subject – a simple call and response. And yet she is wholly convinced that when pigs communicate with each other they do so in a way that is as rich in meaning as it is endless. 'Oh, they are always talking to each other here,' she says breezily. 'It never stops, and that includes the piglets. There is no time when you can't hear them, apart from at night when they're asleep. On farms where they might be kept apart you can still hear them trying to talk to each other, which is another reason why it's so wrong to keep pigs alone.'

## 'I am here'

A pig is constantly switched on to her surroundings. Even when she's flopped out in the sunshine, that snout continues to work. She samples the air, assessing smells as a means of monitoring the world around her. In the same way, her ears never rest. Whether naturally pricked or flopped over her eyes, they flex and twitch as she listens to her fellow pigs grunt and honk, and responds in kind.

'They're contact calling,' says Professor Mike Mendl, in explaining how the group are in constant communication as a means of cohesion. 'It probably serves some sort of function, as if to say "I am here" and "Everything is OK". In particular, you might hear this when a group of pigs are moving through a wood as they forage. It could indicate that they're happy and enjoying the activity, or simply stating where they are.'

As soon as the Professor explains this to me, I begin to think about pigs in an entirely different light. I had always assumed that the low-level grunts and honks that accompany their activities were meaningless. To my ear, it was just noise for its own sake as a pig loses itself in a world of digging, foraging, dozing and eating. Now, it seemed more like a complex social network based on sounds. With every member contributing updates, each pig is informed of individual activities, their whereabouts and the status of the group as a whole.

In the same way that birds flock together or fish swim in schools, every pig in a group is closely connected to the other. The pattern might not be as apparent, graceful or striking as the murmuration of starlings, but at the core of the porcine collective is the same intensity of communication that goes beyond our understanding. What might seem like a basic oink could be rich in information, determined by frequency, tone, pitch and volume. As Professor

Mendl suggests, a group of pigs is known as a sounder for good reason.

'We can't be sure if their calls might reflect being in a positive state, or if they're informing each other or the boss, or something else,' he points out when we talk about research into the number of different vocalisations a pig might possess. 'Nobody knows for certain, but if a group member becomes alarmed, it will often let out a bark sound, which seems to alert the others that there may be danger around.'

## Sound and fury

To human ears, there are few animal noises more urgent and unsettling than a pig's squeal. It's savage, monstrous, sometimes deafening, and can draw out the hairs on the back of the neck. Such was the intensity of her delivery that in order to employ her vocal chords, Roxi's flanks would contract like some demonic living accordion. In broad terms, pigs squeal when they're alarmed, in pain, agitated or even excited. It could be anything from a cry for help, an expression of joy, or a defence mechanism by a largely defenceless creature.

Whatever the true meaning, the intention of a squeal is clear: that pig wishes to make its presence known.

*I am here. Notice me.*

## Where there is hope …

The provision of shelter, food and water is all part of the deal when it comes to being a domesticated animal. In general, a pig in an enclosure can mark out its day by meal-times. Technically, there is no need to summon food, and yet in groups they tend to precede its arrival with an apocalyptic din.

Having experienced this first-hand, it comes as some comfort when I return to the farm buildings with Wendy and she points out what was once a holiday let. 'I used to have to creep out to feed the pigs before they woke the place up,' she tells me, sounding far more good-humoured about it than I ever did. 'Some breeds are noisier than others,' she adds. 'The Tamworths, in particular, never shut up.'

'It would be nice if they just honked a bit when they were hungry,' I grumble. 'My pigs kicked off like they'd just woken up to a murder scene.'

Wendy smiles ruefully.

'I milk my goats first thing in the morning now,' she says, 'but as soon as I open up the doors to where I keep the buckets, the pigs start screaming. They know the feed is stored in there as well, you see?'

'So, do you feed them first?'

Wendy looks at me witheringly. 'By the time I finish milking, the pigs will have given up,' she says. 'They usually last about five minutes, and then think, "Oh, God, she's not coming!"'

'But you always do,' I say, and question why they need to be so vocal when the meal is guaranteed.

'I think they just create purely in hope,' she says.

I dwell on this for a moment. I have never thought of a creature as being inherently pessimistic or optimistic, but Wendy's observation strikes me as spot on. When Butch and Roxi unleashed their dawn chorus it sounded diabolical. Nevertheless, I couldn't fault the lust with which they delivered every squeal. It was never meant in malice, and the pair were always pleased to see me as I sprinted out to silence them.

That squeal will always be capable of curdling the blood, I decide. Unlike a honk, it's hard to read any sense of joy or anticipation in it, but if we can recognise that the pig has a good heart then it makes that racket a little more bearable.

## The silence of the pigs

Two of my children were very young when we had Butch and Roxi. For them, it was quite normal to see two fully-grown pigs join us on the patio in the summertime. For all the challenges that the pigs presented, it was a formative experience for my little ones. They would often toddle down for a visit, and Butch and Roxi were always gracious hosts.

'Just make sure that you shut the gate when you leave,' I would always say before they headed out on an adventure of their own. From my office, I could hear my son and daughter chatting on their way down to the enclosure.

Butch and Roxi would then join in the conversation while I worked. Like me, or anyone else in the company of pigs, my children would talk with them for quite some time. It was lovely to hear, and generally such a gentle background ambience that I would lose myself in writing. Of course, I always registered their return to the house, and then pressed on knowing they were safely inside.

On one occasion only, it was what I *didn't* hear after the children were back in the house that slowly drew me from the keyboard to the window.

I had grown used to the persistent honking. Along with birdsong or a breeze rustling the trees, the sound of two pigs passing the time had become a part of my aural landscape. If they were talking to one another, then that meant they were content. Squealing was a different story, of course, but on some level registering that I couldn't hear them at all set my alarms bells ringing.

On the one hand it was a relief to see that Butch and Roxi hadn't found a way to break through the fence again. It meant I didn't have to devote the rest of my day to tracking down errant livestock who had no movement licence or permission to trash the village. On the other, the sight of my lawn left me with my hands pressed to my head in horror. In the short time that they'd been at work in the garden, having presumably followed my kids through the gate on their way out, the pigs had done their level best to trash it. Instead of neat strips where I had cut the grass that weekend, I found myself looking at random stripes of turf that the pair had uprooted in their exploration of pastures new.

They were only doing what came instinctively to a pig. What struck me was the fact they were operating in complete silence. Now, it could be that Butch and Roxi were just lost in the reverie of the moment, intoxicated by the rich mineral notes of previously untouched sod and soil.

Alternatively, and knowing my pigs, I suspected they were wise to the consequences of communicating their joy at this lucky break. Rather than report to each other on ploughing up my lawn, they struck a pact of silence in order to maximise the opportunity.

It took some persuading on my part to steer them back into the enclosure. As soon as I made my presence known, they responded by finding their voice once more. This time, while I waved my hands uselessly in front of them, I am fairly sure that they were telling each other through a series of stubborn honks to stay strong and simply ignore me.

Finally, following some blatant bribery with blackberries hurriedly picked from the hedgerow, I closed the gate on Butch and Roxi and turned to survey my garden. Whatever they were saying behind my back just then, the pigs were the last things I wanted to hear.

## In a world of their own

There is a particular run I enjoy near home that takes me along the spine of rolling hills. I follow an ancient chalk path, smiling and nodding at passing dog walkers, but largely lost in my own world. It's a beautiful stretch of countryside. On a clear day, it can sometimes seem like I'm closer to the sky than the villages below.

Towards the end of the run, just as a gentle descent begins, the path cuts through the heart of a sprawling pig farm. Arks made from half-round galvanized steel scatter the landscape. Each one is contained in a paddock that is carefully divided by metal fencing. These large sun-baked and snout-turned spaces are populated by pigs across the generations. They're divided into groups, but it's effectively one big community. I can't say what breed I'm looking at. They're a pale shade of pink, with drooping ears and swishing curly tails. The adults are large and long, and while their young vary in stature, it's the very little ones that operate in the highest gear of them all.

On my approach, I watch and listen for them to register me. The pigs seem quite at peace with their surroundings. Whether they are foraging, resting, playing or simply taking the air, I interpret the endless succession of grunts as a measure of their contentment. Normally, I follow the wide path between the fencing and just run straight through. Today, I find myself slowing to a walk. It's good to catch my breath. As I do so, one adult lifts its great head and regards me. Several smaller pigs in the enclosure closest to the alley break from what looks like a friendly tussle to investigate. Watching them crowd into the corner at the mouth of the cut-through, I wonder if they assume I have food. I show them my hands, aware that the nature of the grunting has evolved. It still originates from seemingly random points across the farm, but now it's more widespread.

If this is chatter, the pigs have just found something new to talk about. I can only imagine what they're saying, and wonder if they'll talk to me. Had Emma been here, she would've opened with a penetrating, 'Pig, pig, *PIG*!'. This was a commonplace call among pig-keepers. Wendy used a similarly effective summons, and while it always earned the immediate attention of the pigs, I just can't bring myself to make my presence known so commandingly.

Instead, like an Englishman abroad, I address them in my own language and hope for the best.

'Hi, everyone,' I say. 'How's it going?'

With the little pigs at the fence, I crouch down and greet them individually. While the very young stay close to the adults, the next generation appear to be the most inquisitive. Meantime, some of the larger pigs have paused in their rooting and digging just to face me. I note their ears twitch and flap over their eyes. Even if they can't see me that well, they are listening.

'It's a lovely day,' I say, rising once more to saunter onwards. I take just a few steps before I stop and wait for one large sow, who has begun to plod towards me. Vocally, this one is more forthright than the others. It's a series of short, sharp grunts, and sounds perfectly friendly to me. 'Good morning!'

By the time the pig has reached me, still grunting as I press my palm against the fencing, I find myself in a full-blown conversation. It's as if she answers everything I say in her own way. She even does so with a tip of her head so that she can regard me from under her ears. I look into her eyes just as she looks into mine, and I'm in no doubt that we are communing. Like Wendy, I read what I want from her response, and learn that she's enjoying the day just as I am. It's warm with a light breeze, and we're both up early to make the most of the peace and quiet.

At one point, I even mimic the sow's honk. She responds in kind. Maybe it's encouragement or perhaps she's simply reporting to the others that this guy has lost his mind. Either way, it's a perfectly pleasant exchange.

By the time we part company, the surrounding chatter has subsided somewhat. I look around. Those adult pigs that were watching have returned to the serious business of digging, while the younger ones resume their games without regard for me. As I walk on, following the path through the heart of the pig farm, I like the fact that they are comfortable with my presence. I feel as if I have been assessed, before the word spread across the community that I present no threat or have anything more to offer than passing pleasantries.

Collectively, they have agreed that I can be trusted. Our language systems are entirely alien to each other, and yet pigs still make an effort to communicate with us. When I give this pause for thought, and appreciate how we attempt to connect in this way, it's really quite uplifting. I resume my run with a cheery farewell, and leave the sound of honking behind me.

Minutes later, approaching a busy road towards the foot of the hill, I feel like I have returned to my world. There is a little car park here. Three hikers are just emerging from a hatchback, while a small group of mountain bikers, clad in Lycra and sporting reflective sunglasses, have stopped in preparation for the climb. I raise my hand to acknowledge them all as I run by. Nobody greets me by return. Both the cyclists and the hikers just regard me as being from a different tribe.

# Talking to the animals

I enjoyed my moment with the pigs on the hill much in the same way I used to like talking to Butch and Roxi as I cleared out their sleeping quarters or mended a fence. I would also converse with my old cat in terms of asking how he was, what he'd been up to across the lane and his napping plans for the day. They weren't serious questions, nor did I expect an answer. The cat just represented company, even if he did simply stare at me with unbridled contempt as I spoke. The dogs are never quite so harsh, and though they pay attention to everything I say, they're a little bit like a canine Alexa: primed to activate on just one word, which in this case would be 'walkies'. Until then, they're effectively asleep on the inside.

Conversations with a pig play out very differently.

Unlike almost any other domesticated animal, whether livestock or a pet, the pig contributes to the conversation. They don't just listen, waiting for a cue for food or exercise, they *talk back*. A parrot can be taught to mimic, which is an amazing feat in its own right, and a dog might squeeze out 'sausages' on cue, but this is a creature who listens and responds to us by its own free will and in its own language. A pig will rarely interrupt as you talk by grunting and honking over you. In effect, it can hold a conversation for as long as you like and do it with good manners.

A few days after my hillside run, I go out again on a different trail. This one takes me around the perimeter of sheep fields, and then across a dairy farm on the far side of a fishing lake. As an experiment, I pause in both places to talk to the animals. The sheep simply bolt as soon as I draw breath, and while the cattle are benign, they are much like my kids in that they don't appear to care what I have to say.

The pig, on the other hand, picks up on our words and then replies. We can only guess what it understands or what it's trying to convey, but it's still remarkable. In fact, this goes further than two people who speak different tongues striking up a dialogue in the hope of some understanding. We're talking about two different *species* attempting to bridge that gap. In my mind, that lifts our relationship into a different league.

## Cilla

Wendy takes me across the courtyard towards the lower field. It strikes me as being a challenging terrain to farm for crops. The land is clay-based, boggy with clumps of grass, and rolls steeply into a gulley. Oak trees, which dim and then brighten as clotted clouds float overhead, seam the landscape on the other side of the fold.

This, she tells me, is a playground for pigs and home to one of her oldest companions. 'Cilla is my old retired mother,' she says, 'She's about ten years old, but pigs can live to 18.'

As we navigate the slop, which is so bad I can feel a tug on my wellingtons every time I lift a foot, I focus my attention on a nearby ark. Three pigs are lazily peering out at us. They're lying close to each other with their faces in the daylight. There's room beside them for a fourth pig, who must have clambered out on hearing Wendy's voice from the track. A stocky little kunekune with buttery-coloured bristles and mud up to her belly picks her way towards us. Judging by the way they greet each other, I can't decide whether Cilla or Wendy sounds the most pleased.

'Now she's talking to me,' says Wendy, over a series of deep-chested grunts from the approaching pig. 'That's it, sweetheart! Come and say hello!'

Following my own attempts at conversing with a pig, Wendy has brought me here to show me how it's done. At least that's how I read the situation on listening to the pair. Every time Wendy speaks, the pig meets her gaze and replies in her own way.

'Cilla and I go way back. We've won prizes together, haven't we?'

Judging by the timing and the tone of her honk, Cilla appears to confirm that they have.

'Is she really conversing with you?' I ask Wendy, levelling with her now. 'I mean, seriously?'

She considers this for a moment while scratching behind the pig's ears. Cilla oinks contentedly.

'I think so,' says Wendy, and then returns her attention to the pig. 'She's listening as well as having her say. So I suppose that means we're having a chat. And it's so nice. She wants to be here. She bothered to come up the field to talk to me.'

Again the pig responds as if having registered her words. I glance at Cilla's lady friends in the ark. If Cilla is in fact communicating to them, they seem to be too busy dozing to notice. In the middle distance, at the point where the field tips out of sight towards the gulley, Wendy's dogs have gathered to dig a hole with unbridled enthusiasm. They seem to take turns and then watch with interest. Wendy tells me they spend their days doing this from one field to another. Compared to the diligent excavation skills of a pig, I do wonder if they risk missing whatever it is they're looking for. Cilla certainly pays no heed whatsoever to the dogs or her sisters in the ark. Her sole focus is on talking to Wendy. I think about asking what she's saying, but then realise that's not the point. What counts here is the simple pleasure of the exchange.

'She doesn't see too well, but you only have to look into her eyes to realise there's so much going on in there.' Wendy switches from scratching behind Cilla's ears to rubbing her flanks. By now, Cilla is leaning hard against Wendy's boot. She still acknowledges her voice, but those grunts are increasingly sounding like the cuddle is winning here. 'Eventually,' says Wendy, 'she'll roll over in the mud and fall asleep.'

I laugh, but it also strikes me as a perfectly acceptable way to close any conversation between friends. Not just for a pig, I think to myself, but a human, if nobody else is looking.

# 6
# The Pig's Snout

## Lost treasure

It's believed to be over 2,000 times more sensitive than the human nose, and with more tactile receptors than the human hand. The pig's snout might not be regarded as strikingly pretty, but it's a formidable instrument worthy of our appreciation.

'The snout is a very muscular structure,' says Professor Mendl. 'It's a big rooting disk with a rim they can use to prise up objects … like my drainage cover.'

I feel his pain to a certain extent. In my experience, once a pig has picked up on a compelling scent there is nothing that will stop it from digging up the source. My first taste of this came before Butch and Roxi supersized themselves, in the short time that they lived in a little ark in my office. I had yet to climb the steep learning curve in livestock management, but it dawned on me at that early stage that pigs of any size weren't suited to indoor life.

This was made manifestly evident to me one morning when I found Roxi with her snout jammed between the living-room wall and the radiator. The creak of metal made it clear to me that she was close to popping it free from its mountings.

At the time, she was portable enough for me to pick her up, despite the tantrum, and return her to my office. Later, on inspecting the radiator and figuring we would just have to wait until winter to find out if it still worked, I prodded around the back and dislodged a digestive coated in pet hair, fluff and dust. Practically fossilised, the biscuit looked like it had been dropped down there during the Cold War. To Roxi, however, it had come to represent the sole reason for her existence.

Unsurprisingly, seeing that it practically crumbled to dust in my hand, the biscuit had gone beyond stale. It didn't smell of anything, in fact. At least nothing I could detect.

'There are lots of nerve endings in the pig's snout, and in the nasal cavity, which is also very sensitive.' Professor Mendl has just shown no surprise at my account of the efforts Roxi made to uncover what would surely have been the worst treat in the world. 'In animal welfare research, if we're going to evaluate how important something is to the animal, we ask them to work for it,' he continues. 'This means we increase the amount of work for less reward and see how much effort they're prepared to put into it.'

Immediately, I think of our cat. If I served him a kibble less than a normal mealtime portion, he'd fix me with a stare and then wait for me to correct the error. A pig, I reckon to myself, occupies the opposite end of the fussy spectrum.

'They never seem to give up,' I say.

'I suspect they would stop if something was immovable,' the Professor suggests. 'There has to be a trade-off between desire and energetic expenditure. For example, a pig might be prepared to push through a certain amount of soil to get to a buried acorn. If there's an even more attractive smell, then that pig might push more, but it won't try to get through a solid floor.'

I think about the biscuit once again. It was way beyond its sell-by date and unfit for human consumption. It turned my stomach, and yet to a creature with a more sophisticated olfactory system it had become a valued prize. How must it feel, I wondered, to occupy a world in which smell could seduce and focus the mind so intently? For a pig, the decomposition process must be like the opening of an ancient treasure chest, from inside which aromas are released that it cannot ignore. To be driven by such sensory delights, I decide, can only make life something the pig appreciates way more than we can imagine. It must give rise to an existence, I think to myself, in which nothing is overlooked and everything appreciated.

While it might have caused me some plumbing problems, and with her guiding force in mind, I can't help but feel a little envious of Roxi's lust for something that had been dropped without care and forgotten about entirely.

## The tool of the trade

There is nothing quite like touching a pig's snout with your hand. Unlike a dog's nose, it's not wet, nor do you risk having your wrist licked in the process. If anything, the snout feels like it's made from strong rubber. You might feel warm air on your palm as the pig sniffs and snorts, and you can be sure she'll be reading far more than you from this moment.

Then there's the rim of the snout, which is a crescent of cartilage that's as firm as it is flexible. It is, in effect, a precision digging tool, and one that every pig instinctively knows how to use to their advantage.

'It's very sensitive,' says Professor Mendl, who shares my appreciation of a biological feature capable of fundamentally changing the surrounding landscape. 'A pig can control the pressure it applies to great effect.'

Our hands might be more dextrous for lots of different activities, but we can't beat the snout when it comes to groundwork. This is what the pig is designed to tackle with great effectiveness, and a reason why it's so unsettling to see them on concrete floors.

'Foraging enriches the pig,' says the Professor. 'There isn't much to do in these kinds of pens,' he tells me, 'and there are fewer problems between pigs when they're occupied.'

To demonstrate this, I only have to watch a pig do what comes naturally, and answer the call to forage. Whether it's my lawn, rough land or meadow, she won't simply trash it.

Yes, within a short period of time that space will look like an incident has occurred with a drunk on a rampage in a digger, but the actual process is really quite methodical.

First, the pig will float its snout over the ground. It's an assessment guided by smell, and when something grabs its attention, it'll nudge at the earth as if seeking to wake it up. In this moment, the pig is at its most gentle. It's effectively finding just the right degree of pressure to apply in order to make an incision.

And when it finds that sweet spot, the rim of the pig's snout becomes the leading edge in the dig. I have watched Butch peel a width of turf away with such care and attention that once he'd finished, I could just roll it right back again. Perhaps the only thing missing would be the tips of the grass roots, which can only be some kind of appetiser before the real work begins.

From here on out, the pig begins what is the heart of the dig. The snout might be the primary implement, but this is an animal that will use both head and shoulders to tackle the task at hand, and does so with as much energy as it can muster.

'There are different digging techniques,' Professor Mendl tells me. 'The pig has a very strong and powerful neck, which it often moves upwards or sideways in a shovelling motion. They can also use their teeth to pull or bite through roots,' he adds, which all comes together to create a powerful excavator on four legs.

Observing a pig at work, it's easy to focus on the mess created rather than the progress it's making. Every time that snout plunges deep, it'll come back up with a scoop of soil that could go in any direction. And yet while we tut and shake our heads, the pig is moving one step closer to its goal. Even if that means hours of labour, it'll keep going with the unswerving determination of the metal detectorist.

Part of the pleasure in watching a pig dig, I think, is in not knowing just what it's hoping to find. The pig knows precisely, of course, having acquired all the data it needs by

drawing the air through its nostrils. It's this heightened sense of smell, combined with a snout it can deploy like a wrecking ball or with the exactitude of a surgeon's scalpel, that enables the pig to complete its work without destroying the prize.

'They must be aware of how close they are to something,' suggests Professor Mendl, 'because they become more delicate as things get closer.'

Often their target is a bulb or a root tuber. Then again, I have seen Butch and Roxi unearth discarded tiles or even wartime tins that haven't seen the light of day for more than 70 years. In every case, however, when that pig lays claim to the object of the dig, it does so with a certain reverence. For it has invested time and effort in reaching this moment, and conjured something tangible from the realm of the senses.

From a human perspective, there is no way that we could achieve the same thing without a range of equipment and level of conviction that's hard to match. Yes, the pig will drag its find from its moorings if needs be. If it's remotely edible – and I include house bricks here – it'll chomp and grind it into a pulp. While the end result may not be pretty, it's the act of getting to this moment that we must recognise as the work of a master in an art we cannot hope to match.

# Beyond the boundary

If there is one thing that a pig will do anything to reach, it's the acorn. This simple nut from the oak tree, which grows inside a rough tough shell and then drops to the ground in autumn, can serve as an aromatic siren call to both the sow and the boar. The acorn might be odourless to our noses, and not just bitter to taste but potentially toxic, and yet to the pig it's so irresistible that it might as well have been dropped by angels.

'I lose my pigs when the acorns fall,' Wendy tells me. We've just said goodbye to Cilla, who has accompanied us to the gate. This affectionate kunekune is still grunting away at her keeper, like an old gossip on borrowed time. Wendy and I lean on the gate overlooking the field we've just left and the climb beyond the fold. 'They go AWOL,' she continues, and gestures towards the trees on the slope. 'Normally, they won't go further than the ditch as it's full of water and steep on the other side, but come autumn, the smell of the acorns draws them. So, once they've finished with my oak, they'll make the crossing. It's quite difficult for them, but they always manage it.'

I consider the oak trees on the slope and up along the ridge. They're some distance away. A kilometre, perhaps, and much of that is a climb. Returning my attention to Cilla, who is now inspecting the ground, I can't imagine how it must feel to pick up on the scent of something that far off and become so entranced by it. Wendy tells me that Cilla's acorn-roving days are over. Nevertheless, she talks about her younger pigs as if they're benign roving gangs, quite literally intent on pushing the boundaries.

'How long are they on that side of the bank?' I ask.

'Sometimes they can be gone for a couple of days,' she says, in a way that makes me think Wendy's as relaxed about it as I would be flustered.

'They can be at it for up to 18 hours at a time, unless they're completely stuffed, and then sleep for six,' she tells me. 'In the evenings, I can usually see them out for the count under the eaves of the trees. They might return to the ditch for a drink, but eventually I have to bring them all the way back.'

The way Wendy describes things makes me think this has become some kind of annual event she enjoys as much as the pigs. I admire her attitude to the pilgrimage they make to worship at the altar of the oaks. For her, it's a lovely way to mark the progress of each year, and a rite of passage for each new generation of her pigs that must see them come home enriched by the experience.

## Before the drop

At the tail end of one summer, a year into their residency with us, Butch and Roxi began to behave quite strangely. Usually, whenever I glanced from the window to check on them, I would see the pair practically open-mine casting in my back garden. Over the course of a few weeks, I would increasingly find them standing stock-still, however, as if posing for a portrait. Even when I went out to check on them, I would have to swing open the gate before they registered my presence.

'Is everything OK?' I asked one time, because frankly, as a novice pig-keeper, I was beginning to worry that they were suffering from some kind of paralysis.

Butch and Roxi did at least respond, and I just had to accept that their low grunts meant my fears were unfounded.

A few days passed before I saw the bigger picture. It began when I realised they were spending most of the time under the eaves hanging over from next door's oak. A mature tree, it towered up from behind the fence and formed a canopy that mottled the sun when it fell upon one side of the enclosure. It also dropped a lot of acorns, which didn't last long on the ground. I'd find Butch and Roxi happily shunting their snouts through the leaves or munching on a find, but then occasionally they appeared to lose all interest. This was when they'd make like statues and go weirdly silent. As I happened to be out there one time when they froze, I also stopped what I was doing to see what I was missing. I set the broom against the fence and listened, hearing only a breeze through the leaves. If I had a snout, I would've sniffed as they were. As it was, all I could do was watch and wait.

Then, with a sigh of wind through the branches, an acorn dropped from above. It had barely hit the earth before Butch was onto it, and when another fell, Roxi followed suit. I observed them doing this quite a bit over the weeks that followed, and am pretty sure I saw Butch catch one in his mouth like popcorn.

When I share the story with Professor Mendl, he tells me it's another example of the pig's ability to learn. Certainly, they were aware that the tree contained treats, but how did they know when one was about to drop? Could they detect a subtle change in the acorn's smell that told them it was ripe, or was it the rustling of the leaves that informed them of an impending fall? Whatever the case, with Butch and Roxi in residence I had no need to reach for my rake that autumn. Once they'd finished off the acorns, they took care of the fallen leaves as well.

## After the rain

If I run before work, I head out as the light breaks. It's a quiet and contemplative time, and if rain has fallen overnight, it can feel refreshing in all sorts of ways. I follow a path through the woodland behind us, catching glimpses of deer and wild rabbit, and then out and around a swathe of meadowland near the river. On my way towards the lane that will take me on a long loop home, I pass a little set-up with three pigs. It's a lovely, private world for this trio. I've never seen the owner, but whoever it is must be up before me because the pigs are often chomping down their breakfast when I see them.

Things are only different in the wake of a downpour. Then, when the early morning air is fresh and earthy, the pigs ignore their trough completely. Instead, leaving their feed untouched, I find them hard at work in one of the two paddocks they occupy on rotation every few months. Just before the changeover, as I find it now, the two areas are in direct contrast to one another. Grasses and weeds have sprung from the recovering land, while the bare soil in the old paddock has been turned and heaped repeatedly. Nevertheless, the pigs attack the old ground with the same vigour and enthusiasm as they might when finally let loose in the paddock behind the dividing fence.

There is a word to describe that refreshing smell that arises when rain falls on dry earth, and that is *petrichor*. It's caused when plant oil compounds in wet soil are activated and released into the air. The word originates from Greek, and partly refers to the fluid that courses through the veins of the mythological gods. It's a scent from the ground that even we recognise, and I wonder how intense it must be for the pig. For a moment I watch these three communing with a rejuvenated earth, and then push on, feeling ready to embrace the day.

## No secrets from a pig

With such a keen sense of smell at its disposal, the pig doesn't just use its snout in the search for food but to uncover information about its fellow pigs. It's all about odours, which can help a pig determine everything from the status of a group or an individual.

A pig is also packed with glands, quite literally from front to back end. From its eyes and mouth to its trotters and genitals, each area provides a specific trove of olfactory information on its state of mind, physical health and sexual receptivity. As a result, every last detail around them is picked up by those highly sensitive snouts and then processed in a heartbeat. In effect, all pigs exist in a personal fog of data. Keeping secrets must be close to impossible. This might be an animal that likes to talk, but the truly intimate communication takes place without a grunt or squeal.

I wonder what this means for the pig in the maze, having led the dominant new arrival astray to keep it from the food. I can only think it has to move quickly, for the truth must linger in its wake.

## The orchard next door

'It isn't just for digging,' Wendy confirms as we swap stories about the adventures our pigs undertake when they follow their snouts. 'It can literally go around corners.'

The snout is indeed a miracle of nature. It's the Swiss pocket knife of the porcine world, with a tool for all eventualities, from assessment and initial probing to wholesale excavation. I am mindful of Professor Mendl's point that if a reward requires effort, the pig must first decide if it's worthwhile. Given the multi-purpose potential of its snout, and the fact that it seems pigs consider the effort to be part of the reward, I can't think of many tasks that Butch and Roxi ever declined. My fencing, for example, tended to keep the enclosure contained. Whenever they broke out, which was no easy job, it was down to something compelling in the air.

In the case of the orchard next door, more mature and abundant than my little tree, my pigs found every aspect of their snouts put to the test. Given their sensitivity to smell, I have no doubt that Butch and Roxi were aware of the apples when they were little more than buds. A pig will often appear to test the air, and I'm quite sure that's when something like a Pippin-in-progress would've come to their attention. It certainly directed their digging towards that side of the enclosure, but as I had bolstered the fence panels with hardboard following a previous breakout, I saw no cause for concern. As long as they were happily occupied and safely contained then I figured incoming smells just added to the rich tapestry of their lives.

The fact that I regularly treated them to slices of apples and pears may not have helped them see reason. Both pigs adored this regular treat, and the excitable noises as I approached the gate with a clutch in hand suggested they could smell me coming. What I had to offer them was never enough, however, and so I suppose that aroma from over the fence became all the more alluring.

To be fair to Butch and Roxi, they did hold off for quite a while. Even when our neighbour's apples fell at harvest time, they simply grumbled restlessly. Of course, I had no idea at the time that they were even tempted. As a human, I can't smell fruit from 20 paces. Until, that is, that fruit falls to the ground and proceeds to ferment.

When I first became aware that next door's crop had reached a point of decay that my pigs could not resist, I feared something terrible had happened. They hadn't picked off the overlapping fence slats. With the hardboard in place that really was too much like hard work. Instead, with all the relentless digging, they had undermined two of the fence posts and simply pushed over a six-foot wide panel section. The fact that Roxi was missing told me who was responsible.

What panicked me was the sight of Butch on the ground beside her escape route. He looked like all the bones in his legs had turned to twigs. A quick check confirmed he wasn't dead but sound asleep, which is when it slowly dawned on me that I could hear the sound of an agitated pig nearby. Checking Butch one more time, I noticed that his whiskers were coated in a sticky, fibrous substance. I didn't require a highly tuned nose to tell me that it smelled a little bit like cider. Duly I returned my attention to the broken fencing. Beyond, I noted the gorged remains of many apples past their prime, and realised what had happened here.

Butch was on the floor because he was intoxicated. It looked like he had ventured out and had a good time but failed to quite make it back to bed. Judging by the noise from next door, it sounded like Roxi had perhaps paced herself a little better. Nevertheless, the squeals and grunts didn't seem quite right to me, and so I picked my way over the fallen fence panel to investigate.

I found my sow amid the apple trees, in a standoff with a root that she had exposed to the sunlight. Poised like a bull in a ring, though swaying slightly, she took one look at me and bellowed.

Not only was Roxi drunk, it seemed to me that she had shaped up into an angry one.

It took some persuasion on my part to weave her back over the dividing line, which I achieved using a mint from my pocket and a lot of shoving and pleading. Butch, by this time, had come to his senses, clambered unsteadily to his feet, and together the pair flopped into their sleeping quarters. I didn't see them again for the rest of the day, which gave me enough time to reinstate the fence and top and turf the dig next door.

Maybe I was better prepared the following year, with reinforced fencing and posts, because the pigs made no attempt to reach the apples when that rich and heady smell wafted over. Alternatively, as I like to think, they fell in with Professor's Mendl's view that pigs are quick learners and had vowed to stay on the wagon after one salutary afternoon.

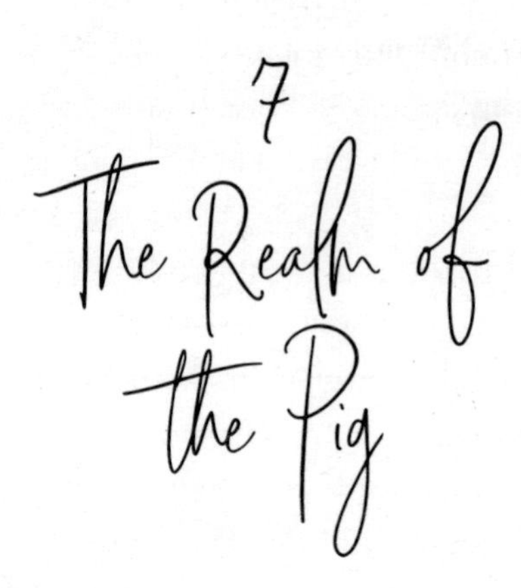


## Come home

'A pig that lives in a wood has got to be the happiest pig of all. It's cool in the summer. It's got everything on the floor that they could ever wish for, and there's a safe feeling in there.'

Wendy tells me this as we continue our tour. We're talking about the kind of habitat that a pig would consider to be just perfect. While she sweeps one hand towards the forested fringes up the hill, I find myself considering a broad and uneven wedge of rough land between two fields. A brook cuts through the clay towards a small copse at the far end, with a cosy shelter nestled on a rise amid tufts of wild grass. If I was a pig, and I have no doubt there are several in the copse or the sleeping quarters who are aware of my presence, this plot would be absolute bliss. Wendy registers me and smiles.

'My pigs mostly go as they please here,' she says. 'In general, it's the sows with litters, or the oldies who don't get into any trouble. But there's so much space for them that they're always going to come home.'

## Home is where the food is

On my run through the forest in Romania, I had no doubt that I was being watched by wild boar. I had entered their realm, where remaining out of sight helped them to feel protected and secure. While the domestic pig has bonded with us, these two factors remain central to their welfare. They have come to trust us, and feel comfortable in our company, but much of that connection is down to the fact that we can provide safe haven. This takes the form of shelter, of course, but there's another factor at play that might well take precedence.

'Did Butch and Roxi come back whenever they ran away?' When Professor Mendl asks me this question, he seems surprised when I tell him that I had to lead them home with a bucket of treats every single time.

'They were just having too much fun,' I tell him, having shared my account of their adventures in my neighbour's orchard and the shaded swathes of ancient bluebells in the woodland across my village. 'I don't suppose they were aware that I could've been prosecuted for it.'

'Well, I'm sure they would have come home when they got hungry,' he offers. 'They always remember a good source of food.'

Naturally, I like to think that having been lured into the wild by some irresistible aroma, my pigs might have sloped back to their enclosure because my family and I represented a good source of love as much as lunch. I tell him that I was unsure Butch and Roxi even knew how to get home, but the Professor is in no doubt. 'Their spatial memory had been closely studied,' he says. 'Pigs will readily return to a site where there's a regular food source.'

According to Professor Mendl, domestic pigs in captivity are known to employ this 'win-stay' strategy, in which they will stick to the same spot if they know they won't go hungry there. At the same time, he points out, when presented with the opportunity they'll opt for what he calls a 'win-shift' strategy. This is effectively old-school foraging, in which pigs will deplete one location of food and then move on to another while mentally mapping it as they go. They can even determine between two food sites, he says, and choose the better option. What brings the loose pigs home, however, rather than behaving like large pink locusts and simply destroying swathes of countryside, has to be the fact that the food on offer is always going to be more plentiful and regular than anything they can find elsewhere.

## The trustworthy pig

As pig-keepers, Wendy and I could not be more different. Where I would go into a state of alarm whenever Butch and Roxi slipped away, Wendy regards a pig with a free pass to be a part of the package. Granted, she lives in a pocket of countyside that is as isolated as it is idyllic. She knows her neighbours, but popping round for a cup of sugar would involve quite a hike. As a result, the sight of a group of sows, along with their sisters and daughters, ambling freely past her farmhouse doesn't send her into the kind of spin I'd have got myself into back home.

'Sometimes they'll go up the track and I won't see them all day,' she tells me quite casually, 'and then at tea time, they'll rock up and go to bed.'

'Does it ever concern you?' I ask, knowing that in her shoes I'd be out in my wellies and wishing I hadn't called my pigs such stupid names as I'm forced to call for them out loud.

She responds with a wry smile. 'There's no need for them to run away from here,' she assures me.

Wendy's calm composure reaches to the four corners of her farm. In listening to her stories about keeping pigs through the years, it's clear to me that very little about their behaviour has caused her to lose sleep. She tells me of instances where a large boar has directed a temper tantrum at her, or fought with another male to the extent that she's had to turn a hose on them to bring the brawl to a close. It's quite evident who's in charge here, and perhaps her pigs recognise that as much as me. It is, I think, a question of trust. In return, they're free to wander because their keeper is schooled in the art of livestock management and quite confident that they'll always come back to a place where life is easy.

# Helga

'Here's a sow and her litter who can go where they like,' says Wendy as we amble towards her courtyard.

There, a low-slung black pig watches over her youngsters. Some share her colouring, others are pink with dark splodges, and all of them look like a test for any mother. They skitter across the concrete, moving in triple time compared to their responsible adult, and squeak like children's toys.

'Who is this?' I ask, mindful not to get too close to a mother and her offspring.

'This is Helga.' Wendy stops at the same distance, but talks warmly to the pig, who responds in kind. 'Her babies are only two weeks old,' she says as several advance under the lowest bar of a gate as if it didn't exist. 'They're going to be lovely, even though Helga is a Swede.' I look across at Wendy, who plunges her hands into the pockets of her overalls and shrugs. 'I bred her with a kunekune to take the edge off the bitchiness.'

For a moment, we watch the little ones at play. It's great to see them gambol freely in the sunshine, and I wonder how far they will roam. No doubt, like any youngsters seeking to find themselves, they'll extend their range in due course. Their mother looks like she's more than capable of shepherding them safely through this formative phase of their lives, however. Her little ones might be a handful, but Wendy recognises that there's no need to keep them penned in. When it comes to setting boundaries, Helga has it covered.

## The pastoral pig

We are so used to thinking that domestic pigs need to be contained. It's often as necessary as it is practical, of course, and with enough land on rotation those pigs lead a happy and fulfilling life. Wendy is fortunate in that her fields have natural boundaries as well as perimeter fencing, but ultimately, the gate at the foot of the long climb to her cottage is wide open. While an intact boar among others requires careful management, of course, she is happy for many in her foraging family to find their own way through each day. In a sense, she makes it work by ensuring that the heart of their world contains all the security they need. In providing food and shelter, she creates a gravitational pull that presents the pig with no reason to move on. But how does she determine, I wonder out loud, whether an individual will play well with others?

'I mix and match based on size, age and aggression,' she tells me. 'It's a healthy thing to do, so long as I understand the pigs and put them on ground that's new to them both, with loads of space. It means a pig can safely run away from a dominant pig,' she explains. 'After a while, the dominant one will give up and eat grass. They get used to each other in this way, and eventually, they'll get tired out. That's when they seal their friendship by lying down and dozing together.'

## Sleeping under the stars

Weather permitting, Wendy's pigs that are entrusted to leave their quarters if they choose are presented with an opportunity for a unique sleepover.

'The heat can change their behaviour quite a bit,' she says as we amble across her courtyard. 'Sometimes on summer evenings they'll head out and find a hedge or a bank and sleep under the stars. On a clear night I can look out and see them cuddled up together in rows.'

Wendy paints a vivid picture for me, and I am completely sold on the idea that pigs would prefer to leave their straw beds when it's stuffy and sleep outside. As she describes once watching a group filing under her window on a hot night and heading out for the cool under the trees, I begin to think of them not so much as livestock but gentle countryside itinerants. I can think of nothing better, in fact, than running at dawn and passing a file of slumbering swine at the boundary between two fields.

While the thought of coming face-to-face with a wild boar instinctively raises my hackles, I have no fear of domestic pigs. I suppose we have lived together for long enough to build into our DNA an inherent respect for one another. They might be formidable in size, but it's no reflection of their temperament. While a mother is naturally protective of her young, and should be treated with caution, encountering a group of sows dozing nose to tail in a ditch in the dew would serve only to brighten my day. While this is unlikely to happen, of course, I appreciate what Wendy has created here simply by stepping back and recognising that many kept pigs can be successfully managed without pens.

## Creature comforts

If there's one thing a domestic pig prizes as much as a regular meal, it's a bed for the night. Collectively, they like to sleep on straw, in a dry space away from drafts, and that's about it. In short, pigs lead a very simple life and this is often reflected in the charming range of sleeping quarters that dot the British landscape.

The pig ark is perhaps the most common kind of dwelling. It's often constructed from galvanised metal sheeting shaped into a half cylinder, with a solid back wall and baffles on each side of the entrance to keep the wind at bay. At the same time, a pig will happily doze in any space provided that meets their basic needs, from stone outbuildings to barns and even shepherd's huts.

When we came to our senses and arranged to house Butch and Roxi outside, I was keen to maximise their living space. One way to achieve this was by customising my garden shed. Rather than plonk a pig ark on perfectly decent foraging ground, I called upon the help of our local pig-keeping friend to assist me in building a chamber at the back of the space reserved for garden tools I never used. With a side entrance and a sloping roof, the space inside easily accommodated both pigs and sheltered them from the elements. Once I'd lined it with straw, I called time on their tenancy inside the house and invited them to follow me.

Ten minutes into the snout-to-surface inspection of the enclosure that followed, both pigs disappeared inside their sleeping quarters. A little bumping and crashing followed, along with a sweep or two of straw onto the ground outside. Then, when a silence fell that was so absolute, I felt the need to investigate.

To be honest, I had worried that after just a short time lounging on sofas inside the house, Butch and Roxi would do nothing but complain about the transfer to a hard floor and no TV. Inside the shed, in case I needed to access the pigs from the top down, we had fixed the roof of the chamber on hinges. Carefully, I cracked it open and peered inside. There, facing the baffled entrance, Butch and Roxi lay side by side and half-submerged in straw. They looked like upturned boat hulls in storage, and were clearly marking their move into the garden with an afternoon nap. Gently, I closed the lid and left them to it.

As the pair settled in, so their snouts began to appear at the entrance during sleeping hours. After a few weeks, Butch and Roxi had taken to dozing with their heads lolling over the threshold. The only variation was if Roxi had clambered inside after a day's digging and then dropped into a slumber without turning. It wasn't unusual, therefore, to find our little boar happily sleeping with his snout jammed between a pair of pink buttocks. It was never the most appealing of sights, but judging by the snores that would accompany this position, I believe they considered it to be a kind of mutual comfort blanket.

## The clean pig

Among their fellow livestock, the pig is perhaps the most misunderstood. From an early age we're led to believe that pigs are messy and unhygienic, but this is far from the truth. As a measure of this, and unlike any other farm animal, it's a fact that pigs will create a toilet area as far from their sleeping quarters as possible. In the house, this cost me the carpet behind the television. Outside, their choice made complete sense, even if it did wake me in the night.

Around three o'clock each morning, I would stir to the kind of exterior thump you might equate with a clumsy burglar falling from the fence. This would be followed by the sound of a grumpy old boar huffing and snorting his way from the shed to the opposite corner of the enclosure. What followed sounded like the garden hose had come alive for 30 seconds, followed by the sound of Butch returning to his quarters. There, another round of crashing and banging would commence, underpinned by complaints from the partner he'd just disturbed, until finally, he settled in again. Inevitably, having been drawn from my own sleep, I would have to get up and go through the same process in our bathroom. Naturally, I would lift the lid first, and I've no doubt that Butch faced the same rigorous standards from Roxi as I did from my wife.

Outside the house – in their rightful environment – the pigs never lived in squalor. Butch and Roxi even undertook the porcine equivalent of changing their bedsheets on a regular basis. Every few weeks, once they'd crushed the straw inside their sleeping quarters to the extent that it lost its spring, the pigs would sweep it out with their snouts. Much of it had been ground to dust, and yet they worked hard to clear the chamber. It also served as a signal to me that I needed to provide them with fresh resources. So I would take this as my cue to go in with a fresh bale. All I ever did was drop it in through the hatch and then lean in to cut the twine.

The first few times I had laid it out nicely, but they just rearranged it and left me feeling like I had done it all wrong. With the bale intact, and before I'd even closed the door at the front of the shed, they'd be inside the chamber, spreading it around to their exact satisfaction. What's more, on regularly shovelling out their latrine area and mixing it with the old straw, I found that my compost heap became a rocket fuel for all the borders in my garden.

Butch and Roxi were content, as was I, and my sunflowers could have won prizes. In working with the pigs and their impeccable standards at each end of the enclosure, we created our own private circle of life. In some ways it countered their groundwork in the middle, which I can only describe as hell on earth.

## The mud myth

Professor Mendl believes that mud is responsible for the greatest misunderstanding of all about pigs.

'If they're kept out in a field, they'll turn it into a swampy mess,' he says, acknowledging that where there are pigs,

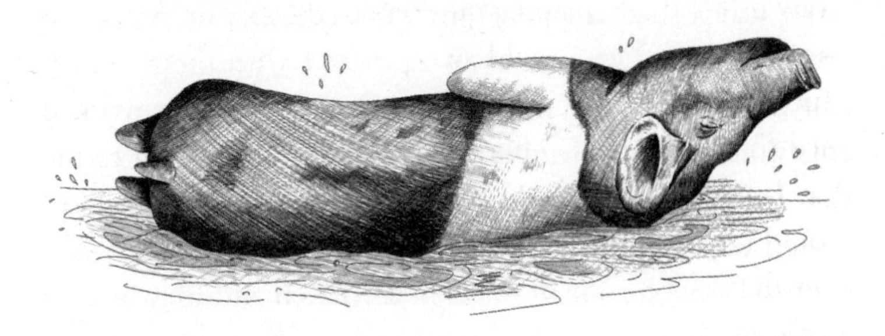

there is muck, but then stresses that this is a by-product of their passion for foraging rather than some in-built appetite for destruction.

Make no mistake, every pig has the potential to churn pristine swathes of greenery into a mire within a short space of time. But it doesn't follow that they either like it that way or are beyond caring.

When I first introduced Butch and Roxi to their new home, it had been previously occupied over several years by chickens. At the time, I thought their persistent scratching at the ground was bad. They had raked away the grass inside the enclosure and turned the surface soil to a fine powder. Within 24 hours of their residency, the pigs had taken things to the next level. Literally. I clearly remember standing at the fence to observe their work. Numb with shock, all I really registered was the fact that they appeared to have uncovered an underground water source beneath the deep strata of clod and clay. The trickle turned out to be run-off from a nearby field rather than a spring I might have bottled and turned to my advantage. Nevertheless, by just adding water to the mix I witnessed the rapid transformation from what was a shaded area of my garden into a recreation of the Somme.

At times, usually after heavy rainfall, the cratered bog became impassable for the pigs in places. They would return from a toilet crossing with a plimsoll line of mud up to their shoulders. Both Butch and Roxi put up with it, and nothing stopped them from their mission to mine around the roots from next door's great oak. Nevertheless, the space was a mess and I sensed that my pigs weren't entirely happy in it.

'We need to let their enclosure rest for a bit,' I suggested to Emma one day during a downpour. Neither pig had even bothered to emerge from their sleeping quarters, and I just felt sorry for them.

'So, where shall we put them?' she asked, and then gave me a look when I glanced through the window at the garden. We had already evacuated the chickens from the enclosure after the ground had threatened to swallow them up.

'Just for a short while,' I said. 'I'll pen off an area and then make it good again afterwards.'

My wife couldn't bring herself to watch later that day when I invited Butch and Roxi to venture onto the lawn. They emerged into the daylight looking fed up, and then pricked up their ears when I directed their attention to the open gate. Beyond, I imagine they saw the lush cut stripes and wondered if they were dreaming. With a kick of her rear trotters, Roxi was first out, and promptly raced around the space with a squeal. Butch took a moment to muster the courage, but quickly followed suit. And as I watched them roll the first of many sods, I knew that was just the price we had to pay for falling for the myth of the minipig. It had seemed like a big sacrifice at the time, but witnessing the upswing in their spirits on driving their snouts into fresh grass made it all worthwhile.

The ground in the enclosure had taken a beating, but after a dry spell and some work reconditioning the soil, I returned the pigs to an enclosure they could turn over again. I also made good on my promise to Emma by turfing the garden where the pigs had trashed it, only for us all to repeat the process during the next round of rainfall that left both areas looking as bad as each other. With heavy hearts, we began to recognise that ultimately, we hadn't just run out of garden but options. It was painfully clear to us both that Butch and Roxi needed more space that could be divided and rotated on a regular basis. We could put up with the noise and the early-morning breakfast commitments, but ultimately, the mess became a measure of their happiness. What mattered here above all else was Butch and Roxi's welfare, which is why we began the unenviable task of seeking pastures new for them.

## Bath time

'People say pigs don't mind mud, but I hate to see a pig wandering through belly-deep. They are very hardy, but I really don't think they like it.'

Wendy is replying to my story of the challenges we faced at the foot of our garden. In a way, I am glad to be on the same wavelength with someone like her when it comes to the issue of pig welfare. Come what may Butch and Roxi would greet each day with gusto. While they showed no sign of discontent, I just wasn't happy with the quality of the ground beneath their trotters.

'The only time they really like it is in the summer,' she continues. 'They don't just need water when it's hot, they actually need mud.'

It's hard to argue that pigs don't like mess when you see one in a wallow. The sight of what appears to be a mud monster blissfully flopping around does undermine the argument somewhat. In this case, however, the action serves a vital function.

'Pigs have no sweat glands apart from the snout,' explains Professor Mendl. 'They'll take advantage of wet mud and coat themselves in it because the moisture evaporates slowly, which helps them to keep cool.'

As well as temperature regulation, it's believed that pigs also turn to mud as a means of lice and parasite control. Having watched my pigs flop into the slop, I also buy into the theory that a nice mud bath is good for their wellbeing. Since Roman times we have coated ourselves in the stuff as a means of healing skin conditions and easing joint pains. While pigs don't pay good money for the same treatment, I like to think they appreciate the way it can help to relax the mind and soothe the soul. When I suggest this to Wendy, she doesn't laugh me off her land.

'My pigs go down into the ditch and make a pool when it's warm,' she says. 'They'll just keep digging and tipping water to create a mess, but it's important to them.'

'Butch and Roxi used to stamp on their water bowl until it flipped,' I tell her. 'I never had a chance to create a wallow for them with a hosepipe.'

I realise I have just made this sound like a lifelong ambition. Wendy tells me pigs are smart enough to take care of it themselves.

'They're also very clever in the way they coat the important places,' she says. 'Firstly, they'll put their noses in it. Next, they'll cover their face to protect against sunburn. Then they do their bits. They rub their bum and tail in the mud and get it down between their buttocks.'

'Really?'

'Well, it's hairless and the sun gets to it,' she says. 'Whether they're male or female, they'll finish by doing their teats, and then, once they've finished all that, they just flounder like a fish.'

I burst out laughing, but not before Wendy. Watching a pig in a wallow is undoubtedly a comical sight, but also rather joyous. As a foraging animal that can happily dig and root for up to 18 hours a day, it's come to accept that mud is part and parcel of its life. It might not like too much, and yet the pig recognises how soil mixed with water contains properties that are beneficial to its health and wellbeing. It's more effective than a cold bath on a hot day and clearly a lot more fun.

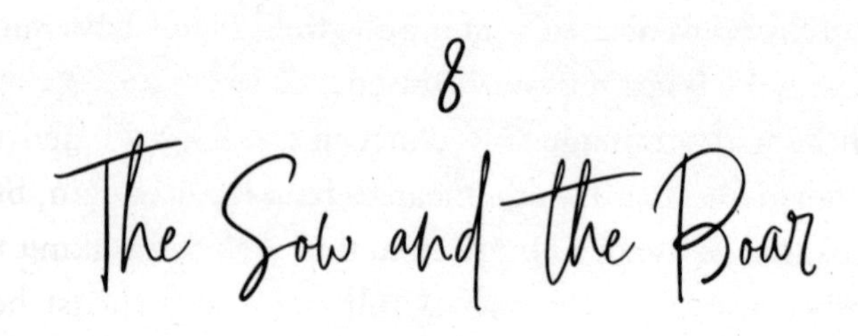

# 8 The Sow and the Boar

## The heat is on

For three weeks at a stretch, Roxi could be a lovely pig. She liked her routine, and whenever I joined her and Butch in the enclosure she would gravitate towards me with a cheery grunt. Yes, she could be boisterous with Butch, who paled in size compared to this huffing pink and splotched-black beast, but essentially she was a sweet-natured sow with a fondness for peaches, an uninhibited attitude to flatulence and a habit of nudging my wellington boots for attention.

And yet, every 21 days, a violent transformation would take place that rivalled Dr Jekyll's switch to Mr Hyde. It could begin at any time, and was marked by a vocal incantation from the enclosure that pretty much silenced the birds in the trees. The first time I heard it from inside the house, I rushed to the window thinking I might need to make an emergency vet callout. Roxi sounded deeply distressed, and yet her squeals were underpinned by something close to fury. When I peered out, I found myself facing a sow who had lifted her front trotters onto the fence as if waiting for my full attention. Whether she saw me or just sensed my presence, she then bellowed at me with such force that the windowpane vibrated.

On hurrying outside, my concern for Roxi turned to bewilderment. She didn't appear to be sick or in pain, but was just bloody angry about something. I tried talking to her. My sow promptly turned full circle and thrust her snout into the soil so aggressively that she caused a volley of soil to break over me. I took a step away, shielding myself from the raining earth, and stared at her in shock. Roxi simply planted her trotters squarely in the ground and glared at me. It was then I noticed that she carried a strong smell with her. I had grown to quite like their natural aroma – a sweet and earthy smell – but at that moment it was overpowering. It was then, as she appeared to give up on me, that I noticed something physically different in my pig. It wasn't just her manner but the folds between her rear legs that had changed beyond recognition.

Just then, drawn by the din, Emma appeared alongside me. I looked at her side-on, and then back at the focus of my attention.

'Is that her vulva?' I asked.

This time, it was my wife's turn to give me a look. It was withering to begin with, but when she glanced back at the pig's rear some concern came into her expression.

'It looks a little swollen,' she said, which was an understatement, and then recoiled with me as Roxi sucked the air into her lungs and screeched.

'Where is Butch?' Emma asked next.

Such was the scene our sow was creating that I had completely forgotten about her companion. I glanced around the enclosure, saw no sign of him, and then registered a pair of eyes peering at me through the straw in their sleeping quarters.

'There,' I said, drawing Emma's attention to the hatch in the side of the shed. 'Hiding under the covers.'

It was our friendly local pig-keeper that informed us that Roxi wasn't dying but simply at the fertile phase in her monthly cycle. It came as a relief, but also quite a shock at such a sea change in her behaviour. Nothing would settle her for several days and nights, which is about the same period Butch spent hiding from her. Then, as rapidly as she came on heat, Roxi returned to her normal casual nature. One minute she was raging at the branches of the oak tree, looking hot and bothered and at war with the world around her, the next she was back to digging contentedly with her soulmate at her side.

When I tell Professor Mendl about what turned out to be a regular event, he has just one question for me: 'Did the boar ever try to mount her?'

'There wasn't much he could've done in that department,' I say, and do that sad snipping gesture that all men seem to understand.

'Well, then she was calling for a mate,' says the Professor.

As he tells me about the profound hormonal changes that the female pig undergoes during the ovulation period, having reached sexual maturity from between three to 12 months of age, I consider whether Roxi was directing all that anger and frustration at me for failing to deliver her a suitor that could meet her needs. The Professor tells me that the agitated state I described, along with pricked ears and a tendency to stand rigidly as an indication of sexual receptivity, is not unusual for a young sow in heat. Still, I can't help wondering if Roxi might have improved her pulling potential had she just calmed down a bit. Then I remind myself that I am interpreting her behaviour through a human filter. Granted, Butch made himself scarce just as soon as she kicked off, but what sounded like tormented snorts and squeals to me was in reality the sound of a pig possessed by the need to perform a basic reproductive function. I imagine an intact boar would've picked up on the change in her perfume and the restless clamouring and reacted like a giant sexy acorn had just materialised especially for him.

According to Professor Mendl, it's a clear window of opportunity that the boar can't afford to ignore. 'If he shows an interest in a sow and she's not in oestrus, she can be quite irritable,' he tells me, which leaves me thinking I did Butch a favour by putting him out to pasture rather than potentially exposing him to even greater wrath when Roxi wasn't in the mood.

## Let's hear it for the boars

As a youngster, staying on the Somerset Levels with my grandparents, I once found myself surrounded by a herd of cattle. I was following a footpath across a field at the time. My grandfather had entrusted me to take his two beloved Labradors for a walk, which was a big deal for an eight-year-old, though I'm sure he had probably dispatched them to look after me.

Until then, cows had never bothered me. I barely registered them as I followed the dogs. Across the field, the herd had spread out so wide they didn't even look like a collective. I don't know what caused them to turn on me, but all of a sudden I found them closing in. First, at a saunter, and then at a trot that quickly turned to thunder.

It was, perhaps, my first experience of running as hard as my legs would carry me. Even the dogs took off for a safe distance before rounding back to flank me to the stile. I'm not sure my grandfather really believed me, and no doubt in retelling the story I escalated the drama so I didn't look cowardly in his eyes. Whatever the case, I have always been wary of crossing fields containing livestock, even though I do it on a regular basis as a runner. I'll keep one eye on the cows, and if there's a bull warning I just pick up my pace and hope for the best. Sheep tend to scatter, and so it's not that bad, but pigs are in a class of their own.

Out running, I have yet to encounter pigs that occupy land with a public right of way. Farmers tend to keep them away from those spaces, simply because pigs can be wary of strangers, behave unpredictably if approached, and if a sow is with her young, regard you as a threat. However, there is a stretch I run every now and then that sometimes takes me across a paddock with a boar. He's kept adjacent to a contained field of sows, which allows him to talk and appreciate their presence. The ground itself is rutted and boggy, which is no surprise given the occupant's favourite pastime, but that's not the reason why I generally avoid that field and choose the long way round.

I have no doubt that part of my reluctance is down to my close encounter with livestock as a boy. I don't want to experience that fear again, in which for one heart-stopping moment I am rooted to the spot in an expanse so big it feels like there's no escape. At the same time, the boar in that field is a big old beast. Everything about his physical make-up looks like it has conspired to unsettle me. From the beady eyes to the prehistoric tusks, that bat-like snout and a squeal that could wake the dead, it's just not an animal I wish to take for granted. Then there's his sheer size. He's all bristles and muscle mass, and was surprisingly fleet-footed the one time I thought I should overcome my anxieties. All he did was turn his great head in my direction and then reposition himself square on. That was enough for me to rethink my decision and leave him well alone.

I have no doubt that boar was probably quite relaxed about runners and ramblers crossing his kingdom. I don't suppose he would've been placed in that field had there been any cause for concern. Nevertheless, there is something incredibly powerful about his aura. He's part beast and part bodyguard, quite possibly there to protect his harem from competition. Whether or not I represent such a threat, I'm simply not prepared to find out. In some ways, he intimidates me more than the bull. While both exist to serve and protect, I just feel that when a bull sees red, he fully submits to instinct. It's a tunnel vision of rage, and I wouldn't care to be in silhouette at the end, but somehow a boar in full flight strikes me as being smarter than that. One look in his eyes will tell you that he's thinking in complex ways just as we do, and I'd be reluctant to discover that he's two steps ahead of me.

## Herbie

The world surrounding Wendy on her farm is profoundly peaceful. From the courtyard across the fields and into the woodland, her pigs are free to dig and play, bask, sleep and contentedly rear their young. Chickens scratch in the dirt, while her three dogs rush from one hole to another as if pursuing some challenge we can never understand. She has worked hard to create a life here, where everybody gets on, and much of this is down to her skill and sensitivity towards managing the boars. Wendy also freely admits there have been times when they have tested her.

'Herbie was a sweetie,' she says of one of her early pigs. 'He was a kunekune, renowned for their temperament, and without doubt one of the kindest and softest of creatures. I used to move him around with a board and stick, but I didn't have to touch him, I just steered him along. One winter morning, we were ambling along when suddenly he turned and whacked me so hard that I fell. I was wearing a ski suit as it was so cold, and he grabbed me by the trouser leg. Then he tried to shake me. Well, I panicked, broke free and ran. From a distance I watched Herbie stomp into the yard, and quickly crossed to shut the gate on him.

'What I really needed to do was get him into his stable to cool off because he was looking very angry, but I was frightened,' she says, sounding pained, as if this is the last emotion any pig-keeper wants to experience. 'So, I collected the board and climbed into the yard with him, and straight away he went for me. I'd parked a trailer and a truck in the yard at the time, and quickly I jumped the draw bar between the two to get away from him. Herbie just shot around the back of the trailer to cut me off.'

Wendy stops there to let me digest the implication of such forward thinking. 'I actually thought he was out to get me,' she says.

'What caused him to turn on you?' I ask.

Wendy looks like she has given this a lot of thought over the years. It's as if this incident reminds her that boars have the potential to be deeply dangerous, even if it is a rare event.

'The only thing I can think is that maybe he could smell another boar on the board,' she says.

We move on through the farm. If Wendy's tale has left me feeling a little cautious about male pigs, she shows no such sign with the next pair that we meet. I can't imagine having to deal with a boar as angry as Herbie, which only heightens my respect for her. Then again, the two she introduces me to don't look like they could fight their way out of a paper bag.

## The lesser boar

Neither size nor age nor swaggering good looks is on the side of these guys, though I'm not sure either of them got the memo. The two whiskery beasts in the stable before me adopt a posture that says don't mess, and that endears me to them more immediately than their genuinely alpha counterparts.

'This one is a quarter Meishan,' says Wendy, as we say hello to one clay-coloured fellow with wrinkled skin and floppy ears. 'He was a bit of a mistake,' she adds, having lowered her voice in admitting what must have been an unusual lapse in her seriously impressive livestock management skills. The part Meishan registers my presence with a baritone grunt. Had I not laid eyes on this lardy chap on little legs, I would've found it quite unsettling, the kind of thing you expect to hear in a crowded pub when you accidentally knock into a biker with a pint in his hand.

Wendy draws my attention to the pig with even more impressive skin folds. This one has a snout that looks like he's been voluntarily chasing brick walls, and he joins in with the huffing and puffing. At the same time, he squares up to us in what looks to me like a stance of fight and not flight. Ignoring the growing air of aggression, Wendy reaches through the gate to ruffle the Meishan under the chin.

'He's in here with his little fat friend,' she says cheerily, which also somehow tells these prickly fellows who's in charge.

In a world in which the matrilineal group has room for just one dominant adult male, where does that leave the beta boars like Butch and the macho duo that fail to follow through with anything but submissive squeaks when Wendy pets them?

'In their natural environments, some boars may never have a harem or father offspring because others are in charge,' says Professor Mendl, which leaves me feeling a bit sad. 'In the context of competing for females, however, they are all very aggressive, as that is their ultimate aim. But otherwise they are fairly docile and certainly not aggressive towards sows.'

This chimes with Butch's relationship with Roxi. Regardless of their genetic background and family line, which remained a mystery to us, our castrated boar figured out his place as her companion. Like any long-standing couple, their character and personalities found a way to complement each other, but it was Butch, I think, who changed to fit with his dominant partner. She quite clearly wore the trousers in their relationship, and could be more emotional if things didn't go her way. In response, Butch set aside any calling to lead the way and learned to be a solid little soulmate for her. Whenever Roxi threw a temper tantrum because she fancied an early supper, he never once got caught up in the moment. Instead, from a safe distance, he remained quite calm and in due course that always brought her back to earth. Of course, they had their squabbles – usually when Roxi woke up to the fact that Butch had stolen a march over her on breakfast – but not once did Butch allow it to become a grudge. He was her rock in the enclosure and her comfort pillow in the sleeping quarters, with no need to play the alpha for a sow that was twice his size.

As a measure of his maleness, Butch began to go prematurely bald on top. He was only two at the time, which might've been quite natural for pigs but certainly seemed harsh to me. As a youngster, his bristles were as dark and glistening as a raven's wing. Within the first 18 months, the first strands of grey began to appear around his eyes and snout, and then shortly after that he started losing it on his crown. On oiling his smooth pate one day, with just a tuft of brush left over each ear, I even wondered if he might feel more confident with a toupee. According to our pig-keeping friend down the lane, Butch was probably just 'blowing his coat', like a dog might shed on a seasonal basis. It was, he believed, a pot-bellied pig thing, so we weren't so concerned when Butch also lost it along the length of his back. While that regrew a short while later, his dome stayed defiantly bristle-free. In body, he became our little old man, but I also think as a look it reinforced the fact that his soul was wise beyond his years.

If I consider his role in their great escapes, it was always Roxi who led the way. Looking back on those episodes, having spent time with Wendy and the Professor, I wonder whether Butch only followed to keep a watchful eye on his temperamental and flighty friend. Apart from the time when the fermented apples got the better of him, I always found him close beside her.

In his own quiet way, I think, as if mindful of his physical shortcomings as much as the sensible head on his shoulders, that put Butch in charge.

## Love is in the air

It's no surprise to learn that for an animal with such acute olfactory senses, smell plays a pivotal role in pig courtship. The smell of a sow is incredibly potent to the boar. It can draw his attention, shape and determine his mood, and dictate his behaviour with predictable consequences. It will even make other sows around attempt to mount her, though she simply shrugs them off.

At the same time, the boar sends out signals of his own. He'll mark his territory with urine, and also chomp, chew and gnash his teeth to produce a foam-like saliva. This is rich in a sex pheromone, and works with the heady notes of his wee to alert the sow and make his intentions known. It may not be subtle, and there's certainly no room for romance, but it works.

On average a sow can produce a litter of up to 12 piglets twice a year. That number can reach 20 in some cases, which makes the pig one of the most productive livestock on the farm. Whether they live in captivity, or come and go as they please under the watchful eye of a keeper like Wendy, any breeding requires close care and attention to prevent being overrun by tiny trotters. This begins by determining if the sow is ready for the boar, but, according to Professor Mendl, not all pigs are as easy to read as my Roxi.

'We can use aerosols to detect oestrus,' he tells me, and describes how such sprays contain a synthetic hormone that mimics the presence of the boar. If the sow gets a whiff and locks her back legs, she's in what's called 'standing heat'. This can also be determined by pressing down on the sow's back. If she doesn't seek to get away, the Professor explains, the time has arrived to bring out the boar.

I cannot pretend that the coupling of a boar and sow is anything other than functional. Pigs pretty much follow a script. I can't say if that makes it joyless, but it is clearly efficient. On meeting, with the boar still frothing at the mouth, and the pair sniffing enthusiastically as they circle and size each other up, the boar will often nudge at the sow's flanks with his snout. This might look like some kind of courtship ritual, and maybe there is something flattering in it for the sow, but it's believed that the boar is physically encouraging the release of eggs ahead of the special moment.

The mount, when it happens, can last for up to half an hour. I once watched a few minutes of such an encounter, alongside Emma, midway through a pig-keeping workshop on a farm. The coupling wasn't part of the programme, but it was a bonus. We had just been wandering around during the lunch break and found them in full swing. Every now and then the boar would thrust as if to remind the sow that he wasn't asleep at the wheel, which I learned is standard pig practice, but mostly he just clung on. The sow, meanwhile, seemed more interested in grazing.

'Priorities,' my wife observed, before we decided to leave them to it.

While I missed the climactic moment, which can last for up to three minutes, I am heartened to learn during my conversation with Wendy that the boar doesn't simply move on in search of his next conquest. They make time for each other, even if the reason why is open to interpretation.

'There's definitely a connection,' she assures me. 'Once they've mated, they will lie down and curl up with each other. It's rather sweet, and cuddly, but the truth is they're probably both knackered. I don't think they're saying, "Ah, that was nice".'